The Wax Moth: A Problem or a Solution?

Authored by

Lovleen Marwaha

*Department of Zoology,
School of Bioengineering and Biosciences,
Lovely Professional University
Phagwara, Punjab, India*

The Wax Moth: A Problem or a Solution?

Author: Lovleen Marwaha

ISBN (Online): 978-981-5123-82-1

ISBN (Print): 978-981-5123-83-8

ISBN (Paperback): 978-981-5123-84-5

need for a court order if at any point you breach any terms of this License Agreement. In no event will any delay or failure by Bentham Science Publishers in enforcing your compliance with this License Agreement constitute a waiver of any of its rights.

3. You acknowledge that you have read this License Agreement, and agree to be bound by its terms and conditions. To the extent that any other terms and conditions presented on any website of Bentham Science Publishers conflict with, or are inconsistent with, the terms and conditions set out in this License Agreement, you acknowledge that the terms and conditions set out in this License Agreement shall prevail.

Bentham Science Publishers Pvt. Ltd.
80 Robinson Road #02-00
Singapore 068898
Singapore
Email: subscriptions@benthamscience.net

BENTHAM SCIENCE

CONTENTS

FOREWORD

Honey bees, next to man, are the most wonderful creation of God. There is nothing that a honey bee does, right from the production of sweet honey to the painful sting it inflicts, which is not in one way or another useful for man. The hive of the honey bee is a treasure vault of valuable products that attract mankind and other animals. No wonder then that the bee hive is under attack from pests, parasites, and enemies. The wax moth is one such problem in apiaries. Belonging to the order Lepidoptera of the class Hexapoda, there are two types of wax moth-infesting honey bee colonies the world over; the greater wax moth Galleria melonella and the lesser wax moth *Achroia grisella*. Old combs, combs in storage and unattended weak colonies are more prone to attacks which can be seen in the form of hard-cocooned pupae on the top of bee frames and anastomosing silken threads of webbing on the surface of the comb. The larvae tunnel through the wax comb, feeding on it and damaging comb and brood in a manner that hinders honey bee life processes ultimately lead to absconding, causing economic loss to the beekeeper. The problem, therefore, needs to be urgently attended and there are chemical, biological, and physical measures for the same. The book addresses all these in a clear and immaculate manner. What is more interesting is the novel concept of utilizing this pest for its more useful activity of plastic degradation. An insect capable of metabolizing complex fatty acids and esters making up beeswax can definitely adapt to digesting similar components that constitute plastic. Whether this is done with the help of gut microbiome as in some cellulose-digesting insects such as termites and cockroaches or by the insect's own enzymes is the unresolved question. This chapter in the book makes it an interesting reading material. Honey bee lovers, parasitologists, entomologists, naturalists, and environmentalists, will definitely enjoy the book. My good wishes to the author.

Neelima R.Kumar
Department of Zoology
Panjab University
Chandigarh. 160014.

PREFACE

The book Wax Moth: A Problem or Solution? has been written especially for B.Sc., M.Sc., and Ph.D. students, bee-keepers, and academicians highlighting various aspects of the wax moth life cycle. Although certain books are available on various pests, pathogens, and predators of the honey bees, very limited information is available on the wax moths. The wax worms are devastating pests of food commodities and the beekeeping industry, which results in great economic loss globally. The wax moth invades the bee colony, proliferates, reproduces, and destroys the colony, forcing the original occupant to abandon the hive. Its biotic potential is regressive, which overpowers the dominance and hold of honey bees in hives. Different control strategies help in dealing with the challenging infestation of a concerned pest, but those also influence apicultural products.

The wax moth possesses the potential to degrade different types of plastics. A few research explorations are available that witness this exceptional ability of the above-cited insect. The present book highlights the specific characteristics and uses of the wax moth to resolve challenges of the human population. For the completion of this book, a vast review of the literature has been carried out, and this book carries the latest information related to the distress pest of honey bees.

During the preparation of this book, authenticated information from various research papers, review articles, and other books has been considered. While practising apiculture, the challenge imposed by wax moths for beekeepers has been observed. This has motivated the compiling of available information about the wax moth in one book. Images incorporated in various chapters are original, having been clicked while maintaining an infested honey bee colony.

Currently, different synthetic chemicals are used that no doubt control wax moth infestation but impose an insecticide contamination problem in apiculture products. On one hand, excessive synthetic chemical application induces insecticide resistance in pests, and on the other hand, these chemicals degrade the nutraceutical value of natural bee products. Therefore, the challenging need for this pest control and the right use of this specific pest acted as triggering factors for the completion of this comprehensive book. The concerned books elaborate on the general introduction, morphology, development, pheromonal profile, mating, reproduction, control, and plastic degradation characteristics of specific pests. With these features, it is hoped that this book will be a complete guide for meeting the needs of the students. A special thanks to Prof. Neelima R. Kumar, Department of Zoology, Panjab University, Chandigarh for providing immensely valuable suggestions while drafting the book.

CONSENT FOR PUBLICATION

Not applicable.

CONFLICT OF INTEREST

The author declares no conflict of interest, financial or otherwise.

ACKNOWLEDGEMENT

Declared none.

Lovleen Marwaha
Department of Zoology
School of Bioengineering and Biosciences
Lovely Professional University
Phagwara, Punjab,
India

CHAPTER 1

An Introductory Review on the Wax Moth; a Devastating Pest of the Honey Bees

Abstract: *Galleria mellonella* L. (Greater Wax Moth) and *Achroia grisella* E (Lesser Wax Moth) are honey bees' most disastrous and economically important pests. Furthermore, in comparison to adults, larvae are a primary destructive stage for honey bee colonies. Voraciously feeding larvae prefer to take bee combs, stored pollen reserves, honey, larval and pupal exuviate, slum gum of the hive, wax capping, natural bee wax, and queen-rearing material containing wax in the storage. Larvae bore the hive, constructing silken tunnels in the colony combs to feed on stored products in the hive. The infested combs become covered with a mass of webbing and faecal matter that results in the condition of *gallariasis*. Weaker, queen-less, poorly managed, less ventilated colonies and abandoned bee hives become easy targets of wax moth infestation.

Further, the strong colonies are also prone to infestation, being a potential host for the heavy growth of this destructive breeder pest. However, after infestation, the bee population of strong colonies declines quickly, and eventually, the hive is destroyed. The present chapter highlights the introduction of the concerned disastrous pest, morphology, development, mating, reproduction, and control. The wax moth is considered a **problem** by apiarists. In contrast, while considering other characteristic features of this insect, the potential ability to degrade plastic of variant types, it provides an excellent **solution** to increasing plastic pollution. Considering both characteristics of this insect, the present book is titled **'Wax Moth a Problem or Solution?**

Keywords: The Wax Moth, Honey Bees, Wax Moth Infestation, Galleria mellonella, Apis mellifera.

INTRODUCTION OF THE WAX MOTH

There is a rapid decline in the honey bee population due to habitat loss, habitat fragmentation, pesticide application, pests, and pathogens. In addition, a wide variety of problems, including wax moths, ants, beetles, robber flies, dragonflies, wasps, cockroaches, mites, praying mantis, death-headed moths, termites, birds, lizards, snakes, frogs, mammals, *etc.*, challenge honey bee colonies, throughout

Lovleen Marwaha
All rights reserved-© 2023 Bentham Science Publishers

the world (Williams, 1978; Caron, 1999; Swamy, 2000; Ellis *et al.*, 2013; Lalita *et al.*, 2018; Vijayakumar *et al.*, 2019).

The common name for a species of moth that invades, attacks, and damages honey bee colonies and apiculture products is wax moth (WM). Other names for the wax moth are wax worm, bee moth, and wax miller. Furthermore, WM is referred to divergent species of moth, including, The Greater Wax Moth (*Galleria mellonella*), The Lesser Wax Moth (*Achroia grisella* Fabricius), The Indian Meal Moth (*Plodia interpunctella* Hubner), The Bumble Bee Wax Moth (*Aphomia sociella* Linnaeus), The Dried Fruit Moth (Vistula spp.), *Ephestia kuhniella* (Zell), *E. cautella* and The Flour Moth (Anagasta kuehniella Zeller) (Paddock, 1918; Okumura. 1966; Wilson and Brewer, 1974; Grant, 1976; Shimanuki, 1980; Kumar, 1996; Williams, 1997; Ellis *et al.*, 2013; Chantawannakul, and de Guzman,2016).

Galleria mellonella L. (*G. mellonella*) and *Achroia grisella* E (*A. grisella*), Vistula spp., *Plodia interpunctela*, *Ephestia kuhniella* are associated with infestation of *Apis cerana*, *Apis mellifera*. *Apis dorsata* and *Apis florea* (Kumar, 1996). Among all predominant species are *Galleria mellonella* (*G. mellonella* L.) and *Achroia grisella* (*A. grisella*), which cause tremendous damage to bee hives. Therefore, almost all Asian honey bee colonies are susceptible to infestation (Adlakha and Sharma, 1975: Brar *et al.*, 1985: Viraktamath, 1989).

The greater wax moth (GWM) is a potentially devastating and economically important pest of honey bees. Even Aristotle (284 BC) mentioned the specific pest in his writings on agriculture (Burges, 1978). *Galleria mellonella* belongs to the order Lepidopteran family Pyralidae. GWM was previously classified as *Galleria cereana* by Fabricius and *Galleria obliquella* by Walker, but it was later reclassified as G. mellonella by Linnaeus. Compared to more incredible wax moths, lesser wax moths are less destructive and less common (Paddock, 1918; Williams, 1997; Ellis *et al.*, 2013; Harding *et al.*, 2013; Kwadha *et al.*, 2017).

The wax moth severely harms the beekeeping industry by damaging both normal and abandoned combs (1a-c). Generally, weak colonies become easy targets for the wax moth attacks (Williams *et al.*, 1976; Kapil and Singh, 1983; Basavarajappa, 2011). Furthermore, improperly cleaned wax combs and queen-less, poorly managed colonies are also prone to infestation. Moreover, colonies become vulnerable to wax moth attacks for various reasons, including malnutrition, disease and widespread mortality due to pesticide poisoning (Pirk, 2016). The wax moths are more pronounced in summer and the rainy season. Therefore, the wax moth infestation is more pronounced in summer and the rainy season.

The general life cycle of this lepidopteron pest includes egg, larval, pupal and adult stages. Pupa hatches to adult form, which stays in the hive for a few days/weeks. Mating of male and female wax moths occurs on nearby trees, and the female returns to the hive to oviposit. The eggs get hatched into larvae under favourable conditions. The emerged larvae move on bee wax comb and get distributed in the hive. While feeding on bee wax, the larvae construct web tunnels for their passage (Kebede *et al.*, 2015).

Numerous reports witness the seasonal prevalence of this pest. Lalita *et al.* (2018) reported that wax moth infestation had increased from April to July. After that, a decline was reported. Other studies have found that the greatest wax moth peak activity occurs from June to November (Ramachandran and Mahadevan, 1951; Brar *et al.*, 1985; Gupta, 1987; Garg and Kashyap, 1998). Furthermore, according to Thakur, 1991, the wax moth population increased from March, reaching a peak in August; after that declined. All honey bee colonies of Asia are prone to wax moth attacks (Adalakha and Sharma, 1975; Brar *et al.*, 1985; Viraktamath, 1989). The concerned pest exhibits diapause during larval and pupal stages (Gulati and Kaushik, 2004).

The information above points out that the wax moth is a difficult challenge for the beekeeping industry. However, some reports attribute its use as an excellent test model for toxicological and developmental explorations and a plastic scavenger to address the issue of accumulating plastic on the planet. According to Ellis *et al.*, 2013, Neumann *et al.*, 2013 and Dietemann *et al.*, 2013, the wax moth is an excellent *in vivo* test model for divergent pathology, entomological, and physiological studies. Additionally, it is also used as a lure for fish bait. The wax moth hosts the hymenopteran species *Pimpla turionellae* and *Itoplectis naranyae*, which are potential pests of various lepidopteran insects. The greater wax moth acts as a factitious host for many larval parasitoids and entomopathogenic nematodes (Ehlers and Shapiro-Ilan, 2005). Further, the wax moths can be used for the biodegradation of different types of plastic and eventually produce ethylene glycol (Bombelli *et al.*, 2017). The wax moth is an excellent experimental model for the study of insect morphology, anatomy, physiology, genomics, proteomics, immunology and reproduction.

THE WAX MOTH AS HONEY BEES' PEST

The wax moths include two predominant types, *Galleria mellonella* (Greater Wax Moth) and *Achroia grisella* (Lesser Wax Moth), which are ubiquitous pests of honey bee colonies (Paddock, 1918; Kumar, 1996; Williams, 1997; Dessalegn, 2001; Gillard, 2009; Ellis *et al.*, 2013; Kwadha *et al.*, 2017). The wax moth is a cosmopolitan and notorious pest of *Apis mellifera* and *Apis cerana Fabricius*, with worldwide distribution. The wax moths cause significant damage in tropical and subtropical regions, resulting in a decline in the honey bee population. The greater wax moth is responsible for heavy economic losses of about 60-70 percent in beekeeping.

Wax moths are nocturnal insects which remain concealed in dark places during the day (Gillard, 2009). The concerned pest generally lives in warm, dark areas with poor ventilation in the hive that is not regularly visited by honey bees (Paddock, 1918; Williams, 1997; Ellis *et al.*, 2013). Wax moths are opportunistic pests that invade weaker colonies with enough unattended wax combs. Wax moth females oviposit eggs in darker and unattended combs (Figs. **1a-c**). After three days, eggs hatch out into the larvae, which destroy hexagonal cells of the hive filled with brood, pollen, and honey (Chase, 1921; Warren and Huddleston, 1962; Smith, 1965; Arbogast *et al.*, 1980; Eischen and Dietz, 1986; 1987; Dessalegn, 2001).

Fig. (1a, b). Comb of honey bees after being heavily infested with the wax moth. The disorganized hexagonal cells are visible with silken webs secreted by the larvae while feeding on the bee wax.

Fig. (1c). The click depicting wax moth larvae and the pupae at the bottom of a bee hive.

The larvae of wax moths burrow the edges of unsealed cells and secrete masses of webs, resulting in *galleriasis*. Furthermore, the larvae destroy the hexagonal cells while feeding on bee wax, making tunnels and spinning silken threads that engulf the entire hive, eventually destroying the complete pack and forcing the honey bee colony to abandon the hive. The wax moths are also responsible for causing the bald brood. As the larvae feed on bee wax and remove the caps of the wax cells, exposing the pupae and other tunnels through wax cells can affect the final moult and result in defective pupa development. The greater wax moth larvae moult for 5-8 moults depending upon environmental cues (Chase, 1921; Smith, 1965; Eischen and Dietz, 1987; Ellis *et al.*, 2013).

The specific insect pest generally prevails in most seasons, either in active or diapause form, but becomes most destructive during the warmer climates (Crane, 2000). Their destruction is not very intense at temperatures below 25°C. The wax moths infest weaker colonies, which results in absconding, and similarly, in stronger colonies, destruction has also been recorded in severe infestation (Kapil and Sihag, 1983 Hanumanthaswamy, 2000; Neumann *et al.*, 2013; Ellis *et al.*, 2013; Dietemann *et al.*, 2013). Further, the infestation is more common in queen-less conditions, starvation, diseases, or temperature variations (Lalita *et al.*, 2018; Komala and Devina, 2020; Shiwani *et al.*, 2020).

The more incredible wax moths feed on honey, pollen, and bee wax under natural conditions, and, in laboratory conditions, they can also be maintained on an artificial diet (Delobel and Laviolette, 1969; Desai *et al.*, 2019).

THE DEVELOPMENT OF THE WAX MOTH

The wax moth, a devastating pest of apiculture, is a holometabolous insect with developmental stages including eggs, larvae, pupae and adults.

The Eggs of The Wax Moth

The fertilized eggs are elliptical with a colour of pearly white in air-unexposed conditions and pink in air-exposed conditions. The eggs are marked externally with diagonally wavy lines. The gravid female wax moth prefers to lay eggs in the cracks and crevices of honey bee hives under natural conditions. The females prefer to oviposit from the late evening to the early morning. The pre-emerged larvae can be easily detected shortly before the hatching and are visible under the egg membrane. According to Swamy *et al.* (2008), eggs of the wax moth are 0.44 ± 0.02 mm long and 0.29 ±0.02 mm broad, with a hatching duration of 8.6 ±0.48 days. Further, the egg-hatching time varies according to the different ecological conditions. Elsawaf (1950) reported that the egg stage needs 9-10 days for hatching.

The Larvae of The Wax Moth

The eggs hatch into pale white coloured larvae. Post-hatching, the larvae move in search of food in the bee combs after perceiving gustatory stimuli from bee combs. The larvae prefer to burrow at the outer edge of the comb, thereby moving toward the midrib of the comb. With further growth, the larval colour changes to a dirty grey. The development of larvae occurs in correlation with the temperature and the food.

Further, the growth rate of larvae occurs comparatively faster if fed on dark and old combs with the brood and the pollen than on fresh combs without the stored pollen. The voraciously feeding larvae destroy the comb faster, especially after the 5th instar. The Greater Wax Moth (GWM) larvae can survive with little food, but under such conditions, comparatively more time is required for the development, and the size of images also remains small. Before moulting, the larvae stop feeding, lose their colour, and shed the exuvium of the head capsules and body separately. Various scientific explorations witness seven instars during WM development (Sehnal, 1966; Anderson and Mignat, 1970). The seven successive larval instars develop in 4.40 ± 0.48, 5.20 ±0.4, 6.5 ±0.67, 7.3 ±0.45, 8.3 ±0.45, 8.4 ±0.66 and 9.2 ±0.4 days, respectively. The total larval duration for GWM is about 49.3 ±1.62 days (Swamy *et al.*, 2008), whereas according to Elsawaf, 1950 and Sehnal 1966, the larval period is completed within 39-62 days and 22-27 days, respectively. The variation in larval phase completion can be due to the quality and quantity of the larval food available.

After hatching, the size of larvae ranges from approximately 1–3 mm in length and 0.12–0.15 mm in diameter, whereas before the pupation, the larval size varies from 25–30 mm in length and 5–7 mm in diameter. At the larval stage, it is challenging to discriminate between males and females due to the absence of sex-specific external morphological characteristics. The larvae of a wax moth are polipod with six thoracic and a pair of segment-wise-abdominal legs from the third to sixth segments. During the early moult, the larvae are cream-white, but the colour changes to dark with successive moults. On the larval head, apical teeth are present, which could be responsible due the destructive nature of larvae. Under the microscope, retractable antennae are present in larval stages (Paddock, 1918; Gulati and Kaushik, 2004; Smith, 1965; Kwadha *et al.*, 2017).

The Prepupa and The Cocoon of The Wax Moth

Before pupation, full-grown larvae search for a suitable site for pupation. In a suitable place within a hive, larvae scrape the wooden surface of the hive slightly before becoming converted into pupae. The last instar larvae shrivel in size and construct a cocoon from silk to get enclosed in the structure. The specific enclosure is referred to as the cocoon, which is elongated, challenging, and leathery. The anterior portion of the cocoon is vast, with an exit hole for the emergence of imago. This hole is closed with thin silken material, and at the time of emergence, it is pushed by an adult. The cocoon length is 22.85 ± 1.95 mm, and the mean breadth is 8.05± 0.92mm, with a pre-pupation duration is 2.1 ± 0.53 days.

The Pupa of the Wax Moth

The pupae in wax moths are obtect type, with extremities fused with the body during the ecdysis. In the pupa, behind the head region, a row of small spines is present, which is extended to the abdominal segment. The pupae's colour changes from white to yellow during the pupation initiation. This, later on, turns from brown to dark brown. The pupation occurs within a spun cocoon covered with faecal pellets and frass with an opening for the exit of the enclosed adult.

The pupa possesses a pair of prominent eyes and antennae (Paddock, 1918; Smith, 1965; Gulati and Kaushik, 2004; Kwadha *et al.*, 2017). The mean length and breadth of the pupae were 13.97 ± 0.58 mm and 4.25±0.29 mm, respectively. The pupation phase is completed within 8.6±0.73 days (Swamy *et al.*, 2008), whereas according to Kannagara, 1940 and Sehnal, 1966, the pupal period is completed within 6.5- 8 days and 8-9 days, respectively. A minor variation in the pupation phase can occur due to changes in climatic factors. It is easy to discriminate between the sexes of the pupa, like that of an adult, as the female pupae are comparatively more extended than the male pupae.

The Adult of the Wax Moth

From the pupae, the adult hatches during the night or late evening hours. Shortly after hatching, the adult moths remain inactive due to the stretching and hardening of their wings. The adults of a wax moth are ashy grey, with 1/3 of the fore wing in bronze colour, whereas the hind wing is uniformly grey.

Depending upon nutritional status, the size of an adult varies proportionally. The larvae feed on dark brown combs, which results in large-sized adults, whereas the larvae feeding on fresh combs result in light-coloured and small-sized adults. The adult antennae have a filiform structure and are longer than the head and the thorax. The adult moth prefers darkness for hiding and rarely gets attracted to the light.

According to Paddock, 1918; Smith, 1965; Gulati and Kaushik, 2004; Kwadha *et al.*, 2017, the adults of a wax moth exhibit sexual dimorphism. The female moth is about 15-20 mm in length, with a 31 mm wingspan and 169 mg in weight. The male moths are comparatively smaller and less dark than the female moths. Further, according to the above-cited researchers, in both sexes, the forewings exhibit varying intensities of the pigmentation that is prominent in the anterior 2/3 of portions. In contrast, in the posterior 1/3 of a wing, a mixture of dark and light pigmentation is present.

In the case of the male moth, the outer margin of the front wing is semi-lunar notched, whereas, in the female, the front wing is smooth and without notches. The wing expansion in the female and male moths is 25.18 ±1.22 mm and 24.6 ± 0.53 mm, respectively. In the female moths, the forward projecting labial palps are present, which give a beak-like appearance to the mouth, whereas in the males, a sharply curved mouth shape is present. The adult wax moths do not feed due to a rudimentary and bifurcated proboscis. In males and females, the antennas are 40-50 segmented in males and 50-60 in females. The leg segments of the greater wax moth include the coxa, trochanter, femur, tibia, pre-tarsus and tarsus (Paddock, 1918; Gulati and Kaushik, 2004; Smith, 1965; Kwadha et al., 2017). The life span of a male moth is 16.4 ± 2.69 days, whereas, in the female moth, it is 6.9 ± 0.7 days (Swamy et al., 2008). According to Elsawaf (1950), the adults' longevity is 21-30 days in males and 8-15 days in females.

The wax moth's life cycle varies from weeks to months according to biotic and abiotic factors (Paddock, 1918; Smith, 1965; Gulati and Kaushik, 2004; Kwadha et al., 2017). The intra-specific factors influencing developmental duration and survival include food and cannibalism. The interspecific factors include parasitoids, honey bees, and the small hive beetle (Paddock, 1918). The abiotic factors, including temperature and relative humidity, influence the development of the wax moth. For the development of the wax moth, the average optimum temperature is 29-33 ^{0}C (Warren, and Huddleston, 1962; Nielsen and Brister, 1979; Williams, 1997). The relative humidity required for wax moth growth is 29-33% (Kwadha and Fombong, 2017). The wax moth can survive in the tropics and subtropical regions (Warren, and Huddleston, 1962; Shimanuki, 1980; Shimanuki, 1981; Williams, 1997).

THE MATING, REPRODUCTION AND LIFE CYCLE OF THE WAX MOTH

The wax moth is a holometabolous insect with four developmental stages: egg, larvae, pupa and adult. The wax moth is a nocturnal insect with peak activity during the 18:00–24:00 hours, the first half of scotophase. Nielsen and Brister (1977) observed that freshly hatched moths move to nearby trees during the onset of scotophase for mating. Afterwards, only females come back to the honey bee colony. Male GWMs display unique mating behaviour by producing acoustic sound from a tympanal organ to stimulate females, and, in response, the female fans her wings and produce low-frequency sound (Spangler, 1988). After that, male wax moths release sex pheromones to attract female moths, ultimately mating (Leyrer, and Monroe, 1973: Spangler, 1988).

Post mating, female moths oviposit in a cluster of 50-150 tiny cracks and crevices of a hive to minimize egg detection and enhance survivability (Shimanuki, 1980: Shimanuki, 1981; Akratanakul, 1987; Williams, 1997; Charriere and Imdorf, 1999: Ellis, and Graham, 2013). The gravid female wax moths prefer a solid colony to a weak colony for a potential host (Williams, 1997). A single female oviposits about 300-600 eggs, with a maximum limit of approximately 1600 in her lifetime (Milum and Geuther, 1935; Mohamed and Copple, 1983; Khanbash and Oshan, 1997). On hatching, the wax moth larvae move from cracks and crevices to the honey bee comb, where larvae destroy the bee comb. The larvae prefer to feed on the honey, pollen, and brood and exhibit aggregation and cannibalism (Nielsen, and Brister, 1979; Williams, 1997). The larvae can be reared on an artificial diet consisting of honey, wax, and cereal products (Kwadha and Fombong, 2017). The wax moth has 8-10 moulting instars, and the last instar spins a silk cocoon (Charriere and Imdorf, 1999). The greater wax moth (GWM) pupae do not feed and are immobile, enclosed within a cocoon. For pupation, about 1-9 weeks is required. The wax moth completes its 4-6 generations in a year (Paddock, 1918; Smith, 1965; Gulati and Kaushik, 2004; Kwadha *et al.*, 2017).

The larval period is extended between 22-60 days on average and can also be upto 100 days (Allegret, 1975; Jyothi and Reddy, 1994; Khanbash and Oshan, 1997), whereas the pupal phase is extended to 60-70 days (Kapil and Sihag, 1983; Jyothi and Reddy, 1993; Brar *et al.*, 1996). The wax moth life cycle is completed within six weeks to 6 months. The development and the metamorphosis are influenced by temperature, relative humidity and diet (Burkett, 1962; Bogus and Cymborowski, 1977; Chauvin and Chauvin, 1985; Kumar, 2000; Gulati and Kaushik, 2004).

NATIONAL AND INTERNATIONAL STATUS OF THE WAX MOTH

The first time, the greater wax moth infestation was reported in the Asian honeybee *Apis cerana*. Afterwards, it spreads to northern Africa, Great Britain, some parts of Europe, Northern America, and New Zealand (Paddock, 1918; Akratanakul, 1987). According to Shimanuki, 1980: and Williams, 1997, this pest is ubiquitously distributed wherever apiculture is practised. The GWM infestation has been confirmed in 27 African countries, 9 Asian countries, 5 North American countries, 3 three Latin American countries, Australia, 10 European countries, 5 Island countries (Paddock, 1918; Katzenelson, 1968; ;Delobel and Laviolette, 1969; Nielsen, and Brister, 1977; Nielsen, and Brister, 1979; Shimanuki, 1980; Kapil and Sihag,1983; Szabo, and Nelson Murray, 1986: 1988; Anderson, 1990; Al-Ghamdi, 1990; Williams, 1997; Corrêa-Marques, *et al.*, 1998; Charriere and Imdorf, 1999; Kato, *et al.*, 1999; Hussein, 2000; Hanumantha swamy, 2000;

Verma and Desh, 2000; Shrestha, and Shrestha, 2000; Swart, *et al.*, 2001; Lebedeva, *et al.*, 2002; Gulati, and Kaushik, 2004; White, 2004; Carroll, 2006; Ehlers and Shapiro-Ilan, 2005; Mondragón, *et al.*, 2005; Topolska, 2008; Al-Chzawi, *et al.*, 2009; Kajobe, *et al.*, 2009; Suwannapong, *et al.*, 2012; Al-Ghamdi, and Nuru, 2013; Rasolofoarivao, *et al.*, 2013; Keshlaf, 2014; Simon-Delso, *et al.*, 2014; El-Niweiri, 2015; Kebede, *et al.*, 2015; Pirk, *et al.*, 2016; Base, 2016: Chantawannakul, *et al.*, 2016 ; Philips, 2016; Sohail, *et al.*, 2017; Hosamani, *et al.*, 2017; Lawal, and Banjo, 2017; Desai, *et al.*, 2019).

CONTROL

The wax moth infestation can be controlled by fumigation with a cotton cloth and tobacco leaf with non-chemical techniques (Popolizio and Pailhe, 1973; Shimanuki and Knox, 1997; Charriere and Imdorf, 1999; Crane, 2000). Additionally, other controls include the use of red fire ants, *Bacillus thuringiensis*, heat and cold treatment, proper storage of old combs and empty boxes and male sterile techniques (Burges and Bailey, 1968; Cantwell *et al.*, 1972; Popolizio and Pailhe, 1973; Burgett and Tremblay, 1979; Goodman *et al.*, 1990; Vandenberg and Shimanuki, 1990; Shimanuki and Knox, 1997; Charriere and Imdorf, 1999; Crane, 2000; Hood *et al.* 2003; Ellis and Hayes, 2009; Hood, 2010).

CONCLUSION

Wax moths are potential pests of the honey bee colony, which generally prefer to infest the weak colonies due to the less guarding force of the colony. If wax moths target a strong colony, they destroy the bee comb, forcing the honey bees to abandon the hive. Further, the wax moth has the potential to degrade different types of plastic. Additionally, the wax moth is used in fish baiting and is an excellent experimental model for studying insect morphology, anatomy, physiology, genomics, proteomics, immunology and reproduction.

REFERENCES

Adlakha, RL, Sharma, OP (1975). *Apis mellifera vs. Apis indica. Gleanings in Bee Culture,* 160.

Akratanakul, P. (1987). Honeybee diseases and enemies in Asia: a practical guide. *Food & Agriculture Org.*

Al-Chzawi, AA, Zaitoun, ST, Shannag, HK (2009). Incidence and geographical distribution of honeybee (*Apis mellifera* L.) pests in Jordan. *Annals of the Entomological Society of France, 45*(3), 305-308.

Al-Ghamdi, A., Nuru, A. (2013). Beekeeping in the Kingdom of Saudi Arabia opportunities and challenges. *Bee World, 90*(3), 54-57.
[http://dx.doi.org/10.1080/0005772X.2013.11417543]

Al-Ghamdi, AA (1990). Survey of honeybee diseases, pests and predators in Saudi Arabia.

Allegret, P. (1975). *Comptes Rendusdes de la Societe de Biologie, 169*(2), 399-03.

Anderson, D.L. (1990). Pests and pathogens of the honeybee (*Apis mellifera* L.) in Fiji. *J. Apic. Res., 29*(1), 53-59.
[http://dx.doi.org/10.1080/00218839.1990.11101197]

Anderson, M.A., Mignot, E.C. (1970). The number of larval instars of the greater wax moth, *Galleria mellonella* (Lepidoptera: Pyralidae), with characters for the identification of instars. *Journal of the Georgia Entomological Society., 5*(2), 65-68.

Arbogast, R.T., Leonard Lecato, G., Van Byrd, R. (1980). External morphology of some eggs of stored-product moths (Lepidoptera pyralidae, gelechiidae, tineidae). *Int. J. Insect Morphol. Embryol., 9*(3), 165-177.
[http://dx.doi.org/10.1016/0020-7322(80)90013-6]

Basavarajappa, S. (2011). Study on the biological constraints of rock bee, *Apis dorsata* F. in southern Karnataka. UGC Major Research Project Report, New Delhi, India, Available from: http://www.nationalbeeunit.com/index.cfm?pageid=207

Bogus, MI, Cymborowski, B (1977). Effect of cooling stress on growth and developmental rhythms in *Galleria mellonella. Newsletter. Biological Sciences Series.*

Brar, H.S., Gatoria, G.S., Jhajj, H.S., Chahal, B.S. (1985). Seasonal infestation of *Galleria mellonella* and population of Vespa Orientalis in *Apis mellifera* apiaries in Punjab. *Indian J. Ecol., 12*(1), 109-112.

Brar, H.S. (1985). *Indian J. Ecol., 12,* 109-112.

Brar, H.S. (1996). *J. Insect Sci. (Ludhiana), 9,* 12-14.

Burges, H.D., Bailey, L. (1968). Control of the greater and lesser wax moths (*Galleria mellonella* and *Achroia grisella*) with *Bacillus thuringiensis. J. Invertebr. Pathol., 11*(2), 184-195.
[http://dx.doi.org/10.1016/0022-2011(68)90148-1] [PMID: 5672009]

Burges, H.D. (1978). Control of wax moths: physical, chemical and biological methods. *Bee World, 59*(4), 129-138.
[http://dx.doi.org/10.1080/0005772X.1978.11097713]

Burges, H.D. (1976). Persistence of *Bacillus thuringiensis* in foundation beeswax and beecomb in beehives for the control of *Galleria mellonella. J. Invertebr. Pathol., 28*(2), 217-222.
[http://dx.doi.org/10.1016/0022-2011(76)90125-7]

Burkett, BN (1962). Temperature block of spinning and development in *galleria-mellonella* (lepidoptera). *American zoologist, 2*(3), 396-396.

Cantwell, G.E., Jay, E.G., Pearman, G.C., Jr (1972). P Thompson JV. Control of the Greater Wax Moth, *Galleria mellonella* (L.), in Comb Honey with Carbon Dioxide: Part I. Laboratory Studies. *Am. Bee J., 112*(8), 302-303.

Caron, D.M., Connor, L.J. (2013). *Honey bee biology and beekeeping..* Wicwas Press.Kalamazoo, MI:

Carroll, T. (2006). *A beginner's guide to beekeeping in Kenya..* Nakuru, Kenya: Baraka Agricultural Training College.

Chantawannakul, P., de Guzman, L.I., Li, J., Williams, G.R. (2016). Parasites, pathogens, and pests of honeybees in Asia. *Apidologie (Celle), 47*(3), 301-324.
[http://dx.doi.org/10.1007/s13592-015-0407-5]

Charriere, J.D., Imdorf, A. (1999). Protection of honeycombs from wax moth damage. *Am. Bee J., 139,* 627-630.

Chase, R.W. (1921). The length of life of the larva of the wax moth, *Galleria mellonella* L., in its different stadia. *Trans. Wis. Acad. Sci. Arts Lett., 20,* 263-267.

Chauvin, G., Chauvin, J. (1985). The influence of relative humidity on larval development and energy content of *Galleria mellonella* (L.) (Lepidoptera: Pyralidae). *J. Stored Prod. Res., 21*(2), 79-82.

[http://dx.doi.org/10.1016/0022-474X(85)90025-6]

Corrêa-Marques, M.H., De Jong, D. (1998). Uncapping of worker bee brood, a component of the hygienic behavior of Africanized honey bees against the mite Varroa jacobsoni Oudemans. *Apidologie (Celle), 29*(3), 283-289.
[http://dx.doi.org/10.1051/apido:19980307]

Crane, E. (2000). Prevention and treatment of diseases and pests of honey bees: The world picture. *New Zealand Beekeeper., 10*, 5-8.

Delobel, B, Laviolette, P. (1969). Breeding of Phryxe caudata Rond. (Larvaevoridae) parasite of Thaumetopoea pityocampa Schiff. on an alternate host *Galleria mellonella. CR Acad. Sci., Paris., 268*, 2436-2438.

Desai, A.V., Siddhapara, M.R., Patel, P.K., Prajapati, A.P. (2019). Biology of greater wax moth, *Galleria mellonella.* On an artificial diet. *J. Exp. Zool. India, 22*(2), 1267-1272.

Dietemann, , Ellis, J.D., Neumann, P. (2013). The coloss bee book, Volume II: standard methods for *Apis mellifera* pest and pathogen research. *Journal of Apicultural Research, 52*(1)
[http://dx.doi.org/10.3896/IBRA.1.52.1.09]

Ehlers, RU, Shapiro-Ilan, D (2005). Forum on safety and regulation. *Nematodes as biocontrol agents, 107-114.*
[http://dx.doi.org/10.1079/9780851990170.0107]

Eischen, F.A., Dietz, A. (1987). Growth and survival of *Galleria mellonella* (Lepidoptera: Pyralidae) larvae fed diets containing honey bee-collected plant resins. *Ann. Entomol. Soc. Am., 80*(1), 74-77.
[http://dx.doi.org/10.1093/aesa/80.1.74]

Eischen, F.A., Rinderer, T.E., Dietz, A. (1986). Nocturnal defensive responses of Africanized and European honey bees to the greater wax moth (*Galleria mellonella* L.). *Anim. Behav., 34*(4), 1070-1077.
[http://dx.doi.org/10.1016/S0003-3472(86)80166-X]

Ellis, A.M., Hayes, G.W. (2009). Assessing the efficacy of a product containing *Bacillus thuringiensis* applied to honey bee (Hymenoptera: Apidae) foundation as a control for *Galleria mellonella* (Lepidoptera: Pyralidae). *J. Entomol. Sci., 44*(2), 158-163.
[http://dx.doi.org/10.18474/0749-8004-44.2.158]

Ellis, J.D., Graham, J.R., Mortensen, A. (2013). Standard methods for wax moth research. *J. Apic. Res., 52*(1), 1-17.
[http://dx.doi.org/10.3896/IBRA.1.52.1.10]

El-Niweiri, M.A.A. (2015). *Survey of the Pests and Diseases of Honeybees in Sudan..* Khartoum, Sudan: UOFK.

El-Sawaf, SK (1950). The Life-history of the Greater Wax-moth (*Galleria mellonella* L.) In Egypt, with Special Reference to the Morphology of the mature Larva (Lepidoptera: Pyralidae). *Bulletin of the Fouad 1er Society of Entomology, 34*, 247-297.

Garg, R., Kashyap, N.P. (1998). *Perspectives in Indian Apiculture* Agro Botanica, Bikaner (Raj.).

Gillard, G. (2009). My Friend, the Wax Moth. *Am. Bee J., 149*(6), 559-562.

Goodman, R.D., Williams, P., Oldroyd, B.P., Hoffman, J. (1990). Studies on the use of phosphine gas to control greater wax moth (*Galleria mellonella*) is stored honey bee comb. *Am. Bee J., 130*(7), 473-477.

Grant, G.G. (1976). Courtship behaviour of a phycitid moth, Vitula edmandsae. *Ann. Entomol. Soc. Am., 69*(3), 445-449.
[http://dx.doi.org/10.1093/aesa/69.3.445]

Gulati, R., Kaushik, H.D. (2004). Enemies of honeybees and their management–a review. *Agric. Rev. (Karnal), 25*(3), 189-200.

Gupta, M. (1987). Wax moth in *Apis mellifera* L. Haryana, India. *Indian Bee J., 49*, 26-27.

Hanumantha Swamy, B.C. (2000). *Natural enemies of honey bees with particular reference to bioecology and management of greater wax moth Galleria mellonella* (Lepidoptera: Pyralidae). , UAS Bangalore.257.

Harding, C.R., Schroeder, G.N., Collins, J.W., Frankel, G. (2013). Use of *Galleria mellonella* as a model organism to study Legionella pneumophila infection. *J. Vis. Exp.,* (81), e50964. [http://dx.doi.org/10.3791/50964] [PMID: 24299965]

Hood, M. (2010). *Wax Moth IPM, Bee Culture magazine, 138*, 9-10.

Hood, W.M., Horton, P.M., McCreadie, J.W. (2003). Field evaluation of the red imported fire ant (Hymenoptera: Formicidae) for controlling wax moths (Lepidoptera: Pyralidae) in stored honey bee comb. *J. Agric. Urban Entomol., 20*(2), 93-103.

Hosamani, V., Hanumantha Swamy, B.C., Kattimani, K.N., Kalibavi, C.M. (2017). Studies on the biology of greater wax moth (*Galleria mellonella* L.). *Int. J. Curr. Microbiol. Appl. Sci., 6*(11), 3811-3815. [http://dx.doi.org/10.20546/ijcmas.2017.611.447]

Hussein, M.H. (2000). Beekeeping in Africa, North, East, North-East and West African countries, Apiacta 1: p 32-48. Online journal]. Available from: http://www. Beekeeping. com/apiacta/beekeeping_africa.htm [Cited Feb 16 2005]. 2000.

Jyothi, JVA, Reddy, C (1994). *Indian Bee J., 56*, 142-144.

Jyothi, J.V.A., Reddy, C. (1993). *C. Indian Bee J., 55*(1/2), 29-35.

Kajobe, R, Agea, JG, Kugonza, DR, Aliona, V, Otim, SA, Rureba, T, Marris, G (2009). National beekeeping calendar, honeybee pest and disease control methods for improved production of honey and other hive products in Uganda. *National Agricultural Research Organisation, Entebbe.*

Kannangara, A.W. (1940). Bee-Keeping-The Wax Moth. *Trop. Agric., 94*(2)

Kapil, R.P., Sihag, R.C. (1983). Wax moth and its control. *Indian Bee Journal, 45*, 47-49.

Kato, M., Shibata, A., Yasui, T., Nagamasu, H. (1999). Impact of introduced honeybees, *Apis mellifera*, upon native bee communities in the Bonin (Ogasawara) Islands. *Popul. Ecol., 41*(2), 217-228. [http://dx.doi.org/10.1007/s101440050025]

Katzenelson, M. (1968). Argentine beekeeping. *Apiacta, 1*, 1-2.

Kebede, E., Redda, Y.T., Hagos, Y., Ababelgu, N.A. (2015). Prevalence of wax moth in the modern hive with colonies in kafta humera. *Animal and Veterinary Sciences, 3*(5), 132-135. [http://dx.doi.org/10.11648/j.avs.20150305.12]

Keshlaf, M. (2014). Beekeeping in libya. *Int. J. Biol. Biomol. Agric. Food Biotechnol. Eng., 8*, 32-35.

Khanbash, M.S., Oshan, H.S. (1997). Arab J. Pl. *Protec., 15*(2), 80-83.

Kumar, Y. (1996). *Advanced Training Course in Apiculture, CAS, Dept.' of Entomology.* CCS Haryana Agric. Univ., Hisar.

Kwadha, C.A., Ong'amo, G.O., Ndegwa, P.N., Raina, S.K., Fombong, A.T. (2017). The biology and control of the greater wax moth, *Galleria mellonella*. *Insects, 8*(2), 61. [http://dx.doi.org/10.3390/insects8020061] [PMID: 28598383]

Kwadha, C.A., Fombong, A.T. (2017). Identification of larval aggregation pheromone components in the greater wax moth, *Galleria mellonella*.

Lalita, Y.K., Yadav, S. (2018). Seasonal incidence of Greater wax moth, *Galleria mellonella* Linnaeus in *Apis mellifera* colonies in the ecological condition of Hisar. *J. Entomol. Zool. Stud., 6*, 790-795.

Lawal, O.A., Banjo, A.D. (2007). A checklist of pests and visitors of *Apis mellifera* adansonii (honeybee) in the six states of south Western Nigeria. *Apiacta, 42*, 39-63.

Lebedeva, K., Vendilo, N., Ponomarev, V., Pletnev, V., Mitroshin, D. (2002). Identification of pheromone of the greater wax moth *Galleria mellonella* from the different regions of Russia. *IOBC WPRS Bull., 25*, 229-

232.

Leyrer, R.L., Monroe, R.E. (1973). Isolation and identification of the scent of the moth, *Galleria mellonella*, and a revaluation of its sex pheromone. *J. Insect Physiol., 19*(11), 2267-2271.
[http://dx.doi.org/10.1016/0022-1910(73)90143-1]

Mondragón, L., Spivak, M., Vandame, R. (2005). A multifactorial study of the resistance of honeybees *Apis mellifera* to the mite *Varroa destructor* over one year in Mexico. *Apidologie (Celle), 36*(3), 345-358.
[http://dx.doi.org/10.1051/apido:2005022]

Murray, R. (1988). Diseases of honeybees in New Zealand. *N. Z. Entomol.,* 15.

Neumann, P., Evans, J.D., Pettis, J.S., Pirk, C.W.W., Schäfer, M.O., Tanner, G., Ellis, J.D. (2013). Standard methods for small hive beetle research. *J. Apic. Res., 52*(4), 1-32.
[http://dx.doi.org/10.3896/IBRA.1.52.4.19]

Nielsen, R.A., Brister, D. (1977). The greater wax moth: Adult behaviour. *Ann. Entomol. Soc. Am., 70*(1), 101-103.
[http://dx.doi.org/10.1093/aesa/70.1.101]

Nielsen, R.A., Brister, C.D. (1979). Greater wax moth: Behavior of larvae. *Ann. Entomol. Soc. Am., 72*(6), 811-815.
[http://dx.doi.org/10.1093/aesa/72.6.811]

Okumura, G.T. (1966). *Bull of the California* Dept. of Agri, California.

Paddock, F.B. (1918). *The beemoth or waxworm.* Available from: http://www.beesfordevelopment.org/media/3245/bfdj-98-news-recent-research.pdf
[http://dx.doi.org/10.5962/bhl.title.57185]

Pirk, C.W.W., Strauss, U., Yusuf, A.A., Démares, F., Human, H. (2016). Honeybee health in Africa—a review. *Apidologie (Celle), 47*(3), 276-300.
[http://dx.doi.org/10.1007/s13592-015-0406-6]

Popolizio, E.R., Pailhe, L.A. (1973). Storing combs in "wax-moth-safe" storage rooms. *Proceedings of the 24th International Apicultural Congress,* 382-383.

Ramachandaran, S., Mahadevan, V. (1951). *Indian Bee J., 13*(7/8), 85-88.

Rasolofoarivao, H., Clémencet, J., Ravaomanarivo, L.H.R., Razafindrazaka, D., Reynaud, B., Delatte, H. (2013). Spread and strain determination of Varroa destructor (Acari: Varroidae) in Madagascar since its first report in 2010. *Exp. Appl. Acarol., 60*(4), 521-530.
[http://dx.doi.org/10.1007/s10493-013-9658-x] [PMID: 23325416]

Sehnal, F. (1966). Critical study of the bionomics and biometrics of the wax moth *Galleria mellonella* reared under different conditions. *Z. Wiss. Zool., 174*(1/2), 53-82.

Shimanuki, H., Knox, D., Furgala, B., Caron, D., Williams, J. (1980). Diseases and pests of honey bees. Beekeeping in the United States. *Agriculture Handbook., 335*, 118-128.

Shimanuki, H., Knox, D.A. (1997). Summary of Control Methods.*Honey Bee Pests, Predators, & Disease* Published by AI Root Company.

Shimanuki, H. (1980). Diseases and pests of honey bees.*Bee Keeping in the United States.* (Vol. Vol. 335, pp. 118-128). Washington, DC, USA: Science and Education Administration, United States Department of Agriculture.

Shrestha, J.B., Shrestha, K.K. (1998). Beekeeping in Nepal: Problems and potentials. Asian bees and beekeeping, Progress of research and development. *Proceeding of fourth Asian Apiculture Association International Conference,* Nepal-Kathmandu262-265.

Simon-Delso, N., San Martin, G., Bruneau, E., Minsart, L.A., Mouret, C., Hautier, L. (2014). Honeybee colony disorder in crop areas: the role of pesticides and viruses. *PLoS One, 9*(7), e103073.
[http://dx.doi.org/10.1371/journal.pone.0103073] [PMID: 25048715]

Smith, T.L. (1965). External morphology of the larva, pupa, and adult of the wax moth, *Galleria mellonella* L. *J. Kans. Entomol. Soc.,* 287-310.

Sohail, M., Aqueel, M.A., Ellis, J.D., Afzal, M., Raza, A.M. (2017). Seasonal abundance of greater wax moths (*Galleria mellonella* L.) in hives of western honey bees (*Apis mellifera* L.) correlates with minimum and maximum ambient temperature. *J. Apic. Res., 56*(4), 416-420.
[http://dx.doi.org/10.1080/00218839.2017.1335824]

Spangler, H.G. (1988). Sound and the moths that infest beehives. *Fla. Entomol., 71*(4), 467-477.
[http://dx.doi.org/10.2307/3495006]

Suwannapong, G, Benbow, ME, Nieh, JC (2012). Biology of Thai honeybees: Natural history and threats.

Swamy, B.C., Rajagopal, D., Kencharaddi, R.N. (2005). Seasonal incidence of greater wax moth, *Galleria mellonella* in Indian Honeybee colonies. *Ind. Bee J., 67*, 3-4.

Swamy, B.H. (2008). Bionomics and biometrics of Greater wax moth *Galleria mellonella* Linnaeus. *Asian Journal of BioScience., 3*(1), 49-51.

Swart, J.D., Johannsmeier, M.F., Tribe, G.D., Kryger, P. (2001). Diseases and pests of honeybees. *Plant Protection Research Institute Handbook,* (14), 198-222.

Szabo, T.I., Nelson, D.L. (1986). *Beekeeping in Western Canada..* Ottawa, Canada: Agriculture Canada.

Topolska, G., Gajda, A., Hartwig, A. (2008). Polish honey bee colony-loss during the winter of 2007/2008. *J. Apic. Sci., 52*(2)

Tremblay, A., Burgeit, M. (1979). Controlled release fumigation of the greater wax moth. *J. Econ. Entomol., 72*(4), 616-617.
[http://dx.doi.org/10.1093/jee/72.4.616]

Vandenbergi, J.D., Shimanuki, H. (1990). Application methods for *Bacillus thuringiensis* used to control larvae of the greater wax moth (Lepidoptera: Pyralidae) on stored beeswax combs. *J. Econ. Entomol., 83*(3), 766-771.
[http://dx.doi.org/10.1093/jee/83.3.766]

Verma, S.C., Desh, R. (2000). Incidence of wax moth (*Galleria mellonella* L. and *Achroia grisella* F.) and their parasitization by Apanteles galleriae wilk. *Himachal Journal of Agricultural Research, 26*, 44-49.

Vijayakumar, K.T., Neethu, T., Shabarishkumar, S., Nayimabanu Taredahalli, M.K., Bhat, N.S., Kuberappa, G.C. (2019). The survey, biology and management of greater wax moth, *Galleria mellonella* L. in Southern Karnataka, India. *J. Entomol. Zool. Stud., 7*(4), 585-592.

Viraktamath, S. (1989). Incidence of greater wax moth *Galleria mellonella* (L.) in three species of honey bees. *Indian Bee Journal., 51*(4), 139-140.

Warren, L.O., Huddleston, P. (1962). Life history of the greater wax moth, *Galleria mellonella* L., in Arkansas. *J. Kans. Entomol. Soc., 35*(1), 212-216.

White, B. (2004). Small hive beetle. *Apiacta, 38*, 295-301.

Williams, J.L. (1997). Insects: Lepidoptera (moths).*Honey bee pests, predators, and diseases.* (pp. 121-141). Ohio, USA: The AI Root Company.

Williams, J.L. (1978). Insects: Lepidoptera (Moths).*Honey bee pests, predators and diseases.* (pp. l05-1127). Ithaca: Cornell University Press.

Williams, JL (1976). Status of the greater wax moths, *Galleria mellonella*, in the united states Beekeeping industry. *Am. Bee. J.,* 524-526.

Wilson, W.T., Brewer, J.W. (1974). Beekeeping in the Rocky Mountain Region. *Colorado State Univ. Cooperative Ext. Service*

CHAPTER 2

Morphometric Characteristics of the Wax Moth

Abstract: The two predominant wax moth species, the greater and the lesser wax moths, exhibit remarkably different morphometric characteristics in the egg, larva, pupa, and adult stages. The eggs of the greater wax moth (GWM) are pink, cream or white, with an ellipsoid, ovoid or obovoid shape, whereas the eggs of the lesser wax moth (LWM) are creamy-white with a spherical shape. Furthermore, in the GWM, the egg size range is $0.44 \pm 0.04 \times 0.36 \pm 0.02$ mm, while in the LWM, the egg size corresponds to $0.41 \pm 0.02 \times 0.31 \pm 0.01$ mm. The first instar larval length in the GWM is 1-3 mm, whereas the last instar body length corresponds to 12-20 mm. In the case of LWM, the first instar of body measurement is 1-20 mm, while in the last instar, it grows upto18.8 ± 0.4 mm. The pupal size in the GWM; is 12-20 mm in length and 5-7 mm in width, while the pupa in the LWM is 11.3 ± 0.4 mm in length and 2.80 ± 1.89 mm in width.

Similarly, in the LWM, the adult body is 10 mm long in the male and 13 mm long in the female moths. The GWM adults possess a 15 mm body length. The dimensions mentioned above for the GWM and the LWM elucidate that the various developmental stages are distinguishable. The present chapter is attributed to the external body dimension and characteristic features of two predominant types of wax moths, which impose significant challenges to apiculture.

Keywords: Wax Moth morphology, Body Size of Wax moth Egg, Wax moth larva, Wax Moth pupa and Adult wax moth, Galleria mellonella.

INTRODUCTION

The wax moth is the common name for many moths that invade, occupy, damage, and eventually destroy infested bee hives. Among all wax moths, the greater wax moth (Lepidoptera: Pyralidae, *Galleria mellonella*) and the lesser wax moth (Lepidoptera: Pyralidae, *Achroia grisella*) are the most destructive pests of the honey bees, which result in considerable economic loss to the bee-keeping industry. The destructive nature of this pest has led to the conduction of numerous studies attributed to life history, biology, behaviour, ecology, molecular biology, physiology, and control. Furthermore, the concerned experimental model is used for divergent explorations dealing with insect genomics, proteomics, insects' early embryonic stages, postembryonic developmental analysis, and pathogenesis studies.

The wax moth undergoes complete ecdysis through four development stages, including egg, larva, pupa and adult, with peculiar body structure, diet type, physiology, proteomics, and developmental and cell differentiation features, as per the coded genomic instructions. The following discussion provides a brief insight into the developmental characteristics of *Galleria mellonella* (G. mellonella) and *Achroia grisella* (A. grisella).

WAX MOTH (WM) EGGS MORPHOMETRIC FEATURE

The eggs of the greater wax moth are white to pink in colour and have a rough texture with diagonally running wavy lines at regular intervals. Further, the eggs are reticulate with a coarse texture and are composed of polygons, including squares, pentagons, hexagons, and heptagons. Although numerous explorations indicate different dimensions and external characteristics of eggs, as indicated by Paddock, 1918; Ellis *et al.,* 2013, the eggs of the GWM are spherical with interspersed wavy lines, with length and width measurements corresponding to the values 0.478 mm and 0.394 mm, respectively. The micropylar area of the egg is surrounded by concentrically arranged microstructure elements similar to rounded flower petals (Swamy, 2008; Ellis *et al.,* 2013; Hosamani *et al.,* 2017; Kwadha *et al.,* 2017; Desai *et al.,* 2019). The micropile is a short minute pore at the cephalic pole of the egg. The egg of the GWM is a flexible structure that usually becomes flat on the sides adjacent to each other and against the substrate to that they are attached. The eggs are externally coated with a film of cement-like substance that enhances cohesion in the form of a cluster.

The surface texture of the eggs of the greater and lesser wax moth differs markedly, which can be considered a peculiar identification feature. The diagnostic difference between the egg of the GWM and the LWM is the feature of reticulation that is faintly visible over the entire egg surface in the GWM, whereas, in the LWM, reticulation is limited to the anterior end (Arbogast *et al.,* 1980; Ferguson, 1987; Williams, 1997; Sharma *et al.,* 2011; Ellis *et al.,* 2013; Hosamani *et al.,* 2017; Kwadha *et al.,* 2017; Desai *et al.,* 2019).

Arbogast *et al.,* 1980; Ferguson, 1987; Williams, 1997; Swamy 2008; Sharma *et al.,* 2011 observed that the GWM egg size ranged from 0.44 ± 0.04 × 0.36 ± 0.02 mm in length and width from 0.29 to 0.39 mm, whereas in LWM egg dimension is 0.41 ± 0.02 × 0.31 ± 0.01 mm in length.

The GWM female deposits eggs in the form of 15-150 egg clusters (Williams,

1997). During about 12 hours of egg hatching, a tiny dark larva is visible through the egg membrane chorion (Paddock, 1918; Williams, 1997). In the lesser wax moth, eggs are oviposited in clusters of 50-150 eggs, each with characteristics of spheroid to ellipsoid, ovoid or obovoid in shape and pink-cream white. During the emergence, the larva bites the eggshell to liberate it.

WAX MOTH (WM) LARVA MORPHOMETRIC FEATURES

The larva of the GWM is polypod, *e.g.* eruciform or caterpillar-shaped, and peripneustic, *e.g.* nine pairs of spiracles type. The body is divided into the head, three thoracics, and 11 abdominal segments (Figs. **2a - c**).

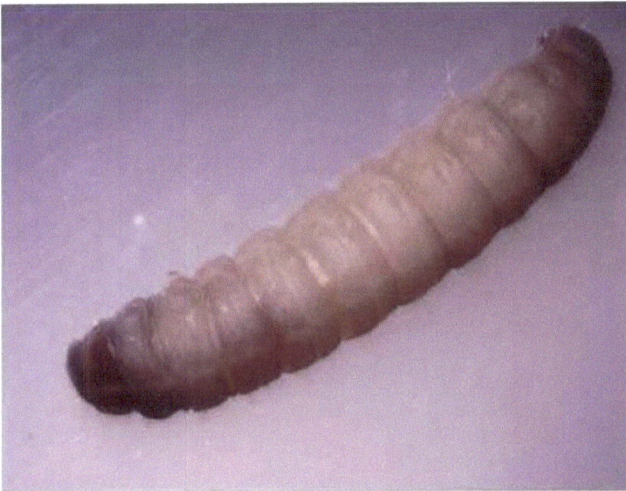

Fig. (2a). Image describing dorsal view of larva of greater wax moth.

The Head Region of the Larva

The highly sclerotised larval head carries a pair of short, two-segmented antennae, chewing mouthparts, and four stemmata on either side (Ellis *et al.,* 2013). The head capsule of the larva is yellowish and comparatively smaller than the prothoracic segments. The larval body colour is grey with a brown prothoracic shield. The head of the larva is slightly pointed, small, and reddish, with a v-shaped line opening towards the front side (Paddock, 1918).

Remarkable external characteristics can differentiate the larvae of the GWM and the LWM. The peculiar identification feature of the GWM larva is the presence of stemmata on the head and spiracles on the larval body. In the GWM larva, the head has four stemmata on each side and spiracles with yellowish peritreme of uniform thickness on the body. In the LWM, stemmata are absent on the larval head, and the spiracle has a thicker black peritreme on the caudal margin. Furthermore, in the GWM larva, the head bears three well-apical teeth but lacks sub-apical teeth. The apical teeth confer an adaptive advantage to the larvae in tearing the bee wax. In addition, the retractable antennae are present, which can be observed under the microscope.

The Thorax and Abdomen of the Larva

In the GWM, the larval thorax is characterised by three pairs of five segmented and a single hook clawed thoracic leg. One oval, brown, and visible spiracles are on the prothoracic segment on either side. In contrast, on the abdominal segments I-VIII, a pair of spiracles are present on either side of each segment. According to Smith 1965, the last larval abdominal segment VIII is the largest of all other segments. In addition, the larvae carry a pair of pro-legs from the third to the sixth abdominal segments.

The first instar larval body is weakly sclerotised and devoid of pigmentation except for the head region. In the final larval instar, the fully covered ecdysial line is opaque toward the mid-dorsal side (Kwadha *et al.*, 2017). Further, the thoracic segments are visible on hatching the larva, but the abdominal segments become remarkable in a three-day-old larva.

The diagnostic differences in colour and larval dimensions are described in the following text.

The Head and Body Colour of The Larva

In the GWM, the larval body colour is creamy-white, grey or dark grey, with a slightly pointed reddish head. While in the LWM, the body is narrow white with brown heads and with clearly visible pronotal shields. After the third instar, the larval body becomes thick and massive due to the voracious feeding habit of the larva (Fasasi and Malaka, 2006; Ellis *et al.*, 2013; Kwadha *et al.*, 2017; Desai *et al.*, 2019).

Larval Body Dimension

Various scientific reports are available in scientific literature attributed to the morphometric marking of the wax moth larva. Arbogast *et al.*, 1980; Ferguson, 1987; Williams, 1997; Sharma *et al.*, 2011; Ellis *et al.*, 2013, the larva of the greater wax moth and the lesser wax moth exhibit differential diagnostic external feature. The GWM first instar larva is 1-3 mm in length and 0.12–0.15 mm in diameter, whereas a fully grown larva possesses a size corresponding to the 12-20 mm range. The LWM first instar larva is 1-20 mm long and 5–7 mm in diameter, whereas the fully grown larva measures a length of about 18.8 ± 0.4 mm.

Desai *et al.*, (2019) described the metric data, including the length (l) and the width (w) of the GWM first to last larva instars as l-0.81, 2.10, 5.86, 8.76, 14.24, 19.58, and 23.88, respectively; w-0.29, 0.44, 1.11, 1.99, 2.03, 2.54, and 3.55, respectively. Further, according to Hosamani *et al.*, 2017, the larval instars (L1-L7) are measured with body length l-1.27, 2.40, 4.80, 9.30, 15.50, 21.60, and 25.40, respectively; w-0.25, 0.45, 1.26, 1.56, 2.65, 3.30, and 4.86, respectively. In the GWM's first to last instar larva, head capsule sizes are 0.21, 0.32, 0.54, 1.15, 1.28, 1.55 and 2.30, respectively. The larvae's morphometric characteristics depend on the food, temperature, and other environmental conditions.

After hatching, the larva moves from the cracks and crevices to the honeycomb, probably stimulated by chemical signals due to bee wax. Paddock (1918) and Nielsen and Brister (1979) showed that the *G. mellonella* larva isolated from the honeycomb always tends to move back toward its food source, highlighting the gustatory signal received by the larva. The wax moth larva feeds on the bee wax by constructing silken tubes or tunnels to destroy the well architecture honeycomb structure. The larva tends to expand its ever-widening tubes toward the central part of the honeycomb, where they tend to aggregate due to involved pheromones. The wax worms also practised cannibalism in the scarcity of sufficient food (Nielsen and Brister 1979; Williams 1997). The larva obtains nourishment from the bee wax for cellular energy production, but the bee wax alone is not a rich protein source (Kwadha *et al.*, 2017). The larval growth is fast if the bee hive contains bee maggots and stored pollen, but larval development is prolonged in a new honeycomb with a little protein source. Therefore, it can be easily assessed that larval phase growth can be accelerated by feeding them a protein-enriched diet. Some explorations conclude that if dietary protein intake is below a certain level, the larva stops spinning the silk. This is probably due to a lack of essential amino acids for silk protein synthesis (Jindra and Sehnal, 1989; Shaik *et al.*, 2017). Further, the protein level also influences larval development. The destruction of comb occurs within a week of colonisation of the specific pest 'WM' (Warren and Huddleston 1962; Nielsen and Brister 1979; Williams 1997; Pastagia and Patel 2007; Swamy 2008; Hosamani *et al.*, 2017; Rahman *et al.*, 2017; Desai *et al.*, 2019).

Larval Molting

The number of larval instars and their growth is determined by food, temperature, and other environmental cues. Different researchers have variable observations related to the number of larval instars. According to Anderson and Mignat 1970; Swamy 2008; Ellis *et al.*, 2013; Venkatesh Hosamani *et al.*, 2017; Desai *et al.*, 2019, the GWM moults about seven times during development, whereas some other rearing experiments even indicated that the wax moth undergoes 8-9 moults during development at 33.8°C (Chase, 1921; Charriere and Imdorf, 1999). Fasasi and Malaka (2006) reported the smallest number of larval instars and concluded that the number depends upon the type of food and other optimal conditions. Nevertheless, authors generally agree with about 5-10 larval instars in the GWM.

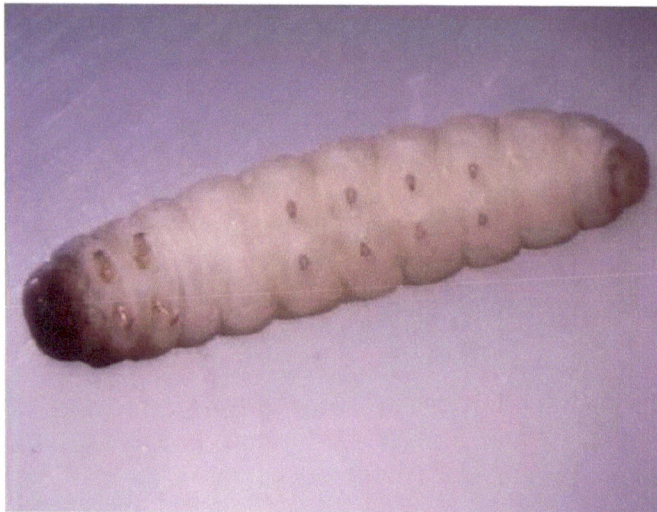

Fig. (2b). Image giving details of ventral view of larva of greater wax moth.

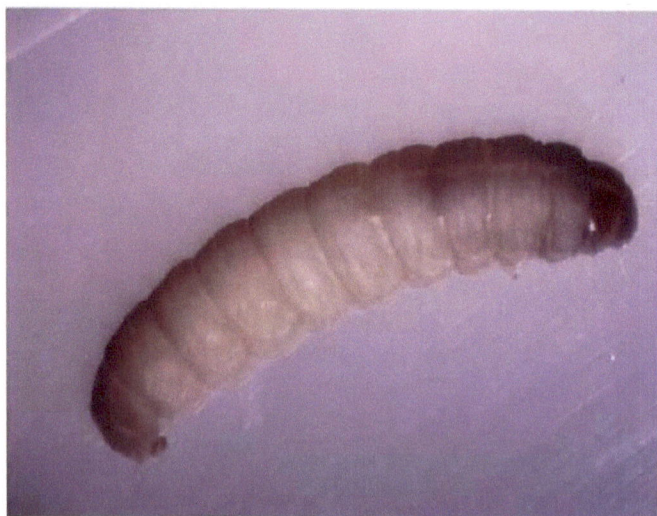

Fig. (2c). Image giving a lateral view of greater wax moth.

The larva suspends feeding before each moulting. During moulting, an old cuticle from the head is shed separately from the rest of the body. The last two larval instar stages exhibit rapid growth if appropriate nutrition, temperature and other conditions are available (Ellis *et al.,* 2013).

The GWM larva spins a protective silken tube which cannot be detected by the honey bees (Shaik *et al.,* 2017). The honey bees can repeatedly be observed removing dead larvae of this pest. The silken tunnels are fabricated from the same type of proteinaceous fibre as that which is used to spin the pupal cocoon (Fedic *et al.,* 2002; Shaik *et al.,* 2017). Sexual demarcation is not possible during the larval phase.

The Laboratory Rearing of the Larva

The wax moth can be reared on wheat flour, corn flour, wheat bran, powdered milk, yeast, honey, and glycerol (Kwadha *et al.,* 2017; Desai *et al.,* 2019). The nutritional status of the larva is correlated with the disease resistance in the larva. The nutritional deficient larva is prone to *Candida albicans* and Berhout infections.

WAX MOTH (WM) COCOON MORPHOMETRIC FEATURES

After finding a suitable place in the hive to pupate, the larva begins spinning silk threads and gets enclosed within it. That composite unit is termed the cocoon (Paddock, 1918). The larvae of the GWM and the LWM exhibit different behaviours for larval aggregation. As in the former-mentioned pest, the larvae tend to aggregate in the hive, whereas in the latter, the larvae hide in tunnels individually within the comb (Williams, 1997). In addition, the larvae excavate in the For cocoon construction, about 2.25 days are required, which is further influenced by the abiotic conditions of the environment (Paddock 1918). In a honey bee hive, the cocoon can be found on the outer surface of bee frames or the inner surface of the hive wall. In the abandoned hive, the pupae can be found anywhere within them.

The slightly shrunken larva becomes inactive for a short duration before undergoing pupation, and the specific stage is referred to as the pre-pupa phase. In the GWM, the pre-pupal degree is not considered a separate developmental stage, as it is not separated from the larval phase by moulting (Chapman 1998). The step during which the larva builds a cocoon and undergoes pupation is the preparatory period (Hosamani *et al.,* 2017; Desai *et al.,* 2019). In the pupal developmental phase, events like histolysis and phagocytosis of the larval structures occur. After that, the imaginary discs form, which comprise the embryonic cells that can divide quickly. These developmental phases are regulated by hormones (Chapman 1998).

In the GWM, the pupae are dark reddish brown enclosed within an off-white-coloured parchment-thick cocoon (Fig. **2d**). At the same time, in the LWM, the colour varies from yellow-tan pupa encircled in a white cocoon often covered in frass and other debris to camouflage in the hive. The GWM pupa is 12-20 mm in length and 5-7 mm in width, whereas the LWM pupa is 11.3 ± 0.4mm, in size and 2.80 ± 1.89 mm, in width. For the pupation phase, approximately 6-55 days are required, which is further dependent on environmental conditions, whereas, in the LWM, about 37.3 ± 1.2 days are necessary for the pupation (Paddock *et al.,* 1918: Smith *et al.,* 1965; Arbogast *et al.*, 1980; Ferguson, 1987; Williams, 1997; Swamy 2008; Sharma *et al.*, 2011; Ellis *et al.,* 2013; Hosamani *et al.,* 2017; Kwadha *et al.*, 2017; Desai *et al.*, 2019). The duration of the pupal phase is dependent upon the temperature and humidity. The pupal phase can be completed within eight days at 28^0C and 65% RH - relative humidity or within 50 days at a temperature ranging from 2.5^0C to 24^0C, and close humidity range from 44% to 100% (Pastagia and Patel 2007; Swamy 2008; Hosamani *et al.*, 2017; Kumar and Khan 2018; Desai *et al.*, 2019).

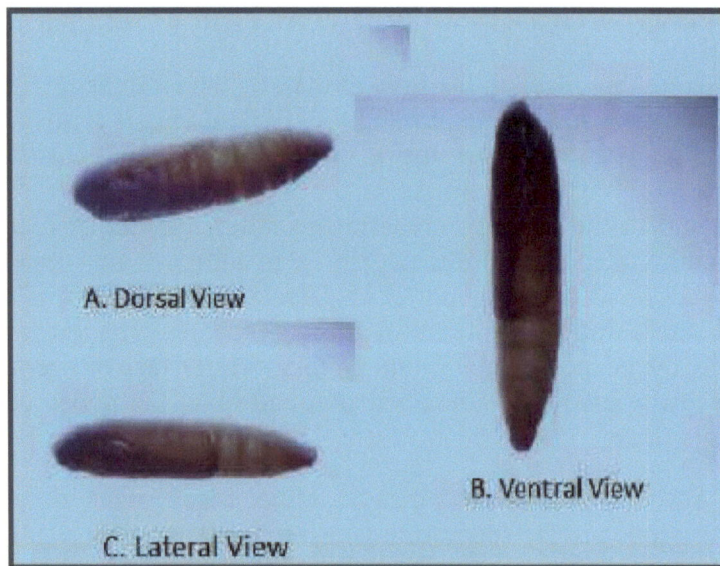

Fig. (2d). Images depicting structural integrity of pupa in dorsal (**A**), ventral (**B**) and lateral view (**C**) in case of the greater wax moth.

The wax moth pupa is the object, meaning all appendages are cemented to the body by a special secretion. A row of spines is present from the head to the 5th abdominal segment (Paddock, 1918).

In the GWM, the sexual differentiation at the pupal stage is as feasible as in the adult stage. The female pupa possesses longer than the male pupae. The female pupa maintains a cloven sternum on the eight abdominal segments for the bursa copulatrix of the male wax moth. The male wax moth possesses a pair of outer round knobs on the ventral side of the ninth (9th) abdominal segment representing the phallomeres. In other words, the ventral sides of abdominal segments VIII and IX exhibit well-marked sexual dimorphism. In the female moth, sclerite VIII is separated, and the IX segment bears single copulatory apertures.

In contrast, in male sclerite of segment VIII is even in appearance, while segment IX bears a pair of rounded knobs representing phallomeres and gonopore between them (Desai *et al.*, 2019). The pupa possesses prominent eyes and antennae. Further, the mesothoracic wings are notched in the male in the GWM but not in the male moth of the LWM. The labial palpi are more significant in females in comparison to males. The pupation occurs within the spun cocoon, covered externally with faecal pellets and frass, to remain hidden in the hive.

The pupal body is elongated with prominent, visible eyes. The antennae of the pupa are long, slightly arched and lengthy up to the second pair of wings. The pre-tarsus of the hind legs protrudes beyond the rear wings' edge (Smith 1965). Two teams of short, protruding setae are present on the parietals, and about 2-7 setae are present on the body segments. Furthermore, in the abdominal segments 2nd to 7th, active spiracles are present on either side of the body.

The cocoons congregate in areas around the perimeter of the bee nest in the high infestations. After hardening the cocoon, the outer layer becomes comparatively solid, whereas the inner layer remains soft. The time for the cocoon formation varies according to the temperature and humidity, with an average value of 2.25 days (Paddock, 1918). There is a reduction in the activity of the larva before the cocoon formation. The larva creates an incision point in the cocoon near the head to escape after the adult forms (Paddock, 1918).

The cocoon protects the larva and pupa from the worker honey bees and other parasites and stabilises abiotic conditions for proper pupal development (Jindra and Sehnal, 1989; Shaik *et al.,* 2017). According to Ellis *et al.,* (2013), the outer surface of the cocoon becomes hard, whereas the inner surface remains soft. In the front area of the cocoon, the larva makes the exit hole for the future adult. Before pupation, this opening is closed off with a thin layer of silk (Paddock 1918; Desai *et al.*, 2019).

Various reports give descriptions of the pupal morphology, including colour and sexual dimorphism (Paddock 1918; Smith 1965; Swamy 2008; Hosamani *et al.*, 2017; Kwadha *et al.*, 2017; Desai *et al.*, 2019). The eclosion of the adult from the cocoon phase occurs during the night or late in the evening. During the emergence, the adult pushes out the silk lids that cover the cocoon for exit holes (Swamy 2008). After the emergence from the cocoon, the adult remains inactive until hardening and full extension of its wings. The moths are creamy white, becoming dark to grey (Nielsen and Brister 1979; Swamy 2008; Desai *et al.,* 2019). The GWM prefers darkness, and they try to conceal themselves at different corners of the hive if exposed to light.

ADULT WAX MOTH (WM) MORPHOMETRIC FEATURE

The male moth is comparatively smaller and lighter than the female moth. In the GWM, the adult size is about 15 mm in length, with a 31 mm average wingspan and 169 mg in weight (Paddock, 1918; Ellis *et al.,* 2013), whereas in the LWM, the male is 10 mm long and the female 13 mm long. In the GWM, adults possess a heavy reddish-brown body with mottled forewings and pale-cream-coloured, lightly fringed hind wings. At the same time, the LWM is a small, silver body with a conspicuously yellow head, oval-shaped forewings and heavily fringed hind wings. The head sclerite and head appendages are covered with heavy coats of scales. In the GWM, the frontal clypeus is not sutured, but a midventral line of demarcation separates the adult head into the right and left halves.

Dorsal view

Fig. (2e). Image of the male greater wax moth adult from the ventral side.

The compound eyes are large and hemispherical, with 0.65 mm in diameter and 4500 to 4,650 ommatidia. Furthermore, each ommatidium measures about 0.0125 mm. In daylight, eyes appear white, but in the dark, eye colour appears glossy black.

The labial palp is a three-segmented structure that is slightly upward and forward in the female, whereas in the male, it is sharply curved upward and hooked inward. In the male, the distal segment has a sharply curved spine. The female moth has a pointed nose, while the male moth has a snub nose.

Both the male and female adult wax moths possess filiform antennae with segment numbers 40-50 in the male and 50-60 in the female (Williams, 1997; Smith, 1965). The antennae possess an enlarged and elongated shape with a bulb at the proximal end, which articulates with the head region. The pedicel is intermediate in size and is present in the space between the scape and the proximal antennal segment. The proximal parts of the antennae are diametrically larger and shorter in size than the distal segment, which is more slender and more prolonged. The ventral side of each antennal segment is covered with tiny tactile hairs or microtrichia. The maxillary and labial palpi are usually developed in the GWM, whereas small vestiges represent the mandibles. The proboscis is underdeveloped and bifurcated in the GWM at the distal end, with the galeae separated distally.

The thorax comprises three segments, prothorax, mesothorax, and metathorax (Figs. **2e** and **f**). In the GWM, mesowings exhibit sexual dimorphism. The distal border is almost straight in females, while in males, it is notched. The prothoracic legs bear no tibial spurs, while the mesothoracic legs bear one pair of large tibial spurs. On the metathoracic wings on the tibia, two pairs of stimuli are present.

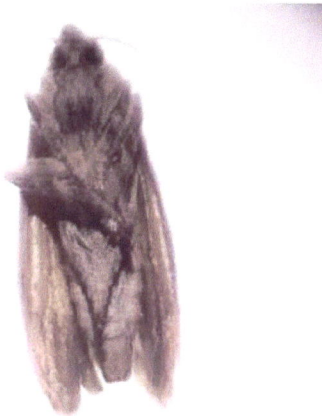

Fig. (2f). Picture of the male greater wax moth highlighting insect appearance from dorsal side.

The forewings in the adult vary in the intensities of pigmentation as the anterior

2/3 of the wing is covered with scales, hence bears uniform dark pigmentation, whereas the posterior 1/3 of the wing is marked with a mixture dark and light pigmentation stripes (Paddock, 1918; Ellis *et al.,* 2013). The male wax moth possesses scalloped forewings, whereas the female moth has straight distal wings. The female moth includes forward projecting labial palps that give a beak-like appearance, while the male moth bears a sharply curved, hooked and inward snubbed nose. The wax moth wings are grey, while the third hind portion of the wing is a bronze colour. The wing venation pattern of the GWM and LWM can be considered a diagnostic feature (Ferguson, 1987). The GWM legs consist of the coxa, trochanter, femur, tibia, pre-tarsus, and tarsus (Kwadha *et al.,* 2017).

Seven pre-genital abdominal segments are present in the GWM male and female moth. The first abdominal segment is significantly modified for articulation with the metathorax. The first pair of spiracles are located in the middle of the first abdominal pleura. Further, a pair of chordotonal organs are present on the first abdominal segment, evidenced by large tympani. In the male, the sternum of the eighth segment is modified into claspers. Male WM segments 9-11 have a combined tergum, whereas female 9-11 segments have been changed to form a telescoping ovipositor. The complete body comprises the legs, wings, and antennae covered by the scales.

In the GWM, the life span is about 12 days in the female, whereas, in the male moth, it is about 21 days. In the LWM, the life span of the female is 6.90 ± 1.135 days, while in the male moth, it is about of 12.90 ± 1.30 days (Arbogast *et al.,* 1980; Ferguson, 1987; Williams, 1997; Sharma *et al.,* 2011; Ellis *et al.,* 2013). The adults do not feed due to atrophied mouth parts and therefore possess a short life cycle, which is further dependent on the ambient conditions (Paddock 1918; ElSawaf 1950; Singh, 1962; Opoosun and Odebiyi 2009; Hosamani *et al.,* 2017; Kumar and Khan 2018). In more incredible wax moths, about 6-55 days are required for normal development in appropriate environmental conditions, whereas in lesser wax moths, about 37.3 ± 1.2 days are required for a normal life cycle.

The adults of the LWM are characterised by their small size of 8 to 13 mm. This species has long, filiform antennae, with the distal segments markedly thinner than the basal segment. The front is covered by orange scales, like hairs. The body is light brown with golden highlights and sprinkled with black scales. The prothorax is covered with black scales. The rounded-contour wings show similar colouration to the body. When in the pre-pupal stage, the LWM moths are found in depressions in the wooden walls of the hives and the combs, weave the cocoons and pupate, usually side by side. Roh *et al.,* 2020 reported a new wax moth species, *Galleria similis*, on the basis of hindwing venation and male genitalia.

CONCLUSION

The wax, a bothersome pest of honey bees, can be easily distinguished by remarkable diagnostic morphometric features of various developmental stages, including egg, larva, pupa, and adult. The specific features include colour, body dimension, developmental duration, and general body integrity. Further, the various developmental stages of the GWM and the LWM can be distinguished from each other with certain visible features that include colour, shape, dimension, body organisation, head, thorax and abdominal appendages, wing venation, spiracle location, genitalia structure, concealing site and behaviour in the larva, pupa, and adults, feeding habit, camouflage, ecological condition preference and their self-protection strategies from the host insect colony.

REFERENCES

Arbogast, R.T., Leonard Lecato, G., Van Byrd, R. (1980). External morphology of some eggs of stored-product moths (Lepidoptera pyralidae, gelechiidae, tineidae). *Int. J. Insect Morphol. Embryol., 9*(3), 165-177.
[http://dx.doi.org/10.1016/0020-7322(80)90013-6]

Chapman, RF, Chapman, RF (1998). *The insects: structure and function.* Cambridge University Press.
[http://dx.doi.org/10.1017/CBO9780511818202]

Charriere, J.D., Imdorf, A. (1999). Protection of honeycombs from wax moth damage. *Am. Bee J., 139*(8), 627-630.

Chase, R.W. (1921). The length of life of the larva of the wax moth, *Galleria mellonella* L., in its different stadia. *Trans. Wis. Acad. Sci. Arts Lett., 20*, 263-267.

Desai, A.V., Siddhapara, M.R., Patel, P.K., Prajapati, A.P. (2019). Biology of greater wax moth, *Galleria mellonella*. On an artificial diet. *J. Exp. Zool. India, 22*(2), 1267-1272.

Ellis, A.M., Hayes, G.W. (2009). Assessing the efficacy of a product containing *Bacillus thuringiensis* applied to honey bee (Hymenoptera: Apidae) foundation as a control for *Galleria mellonella* (Lepidoptera: Pyralidae). *J. Entomol. Sci., 44*(2), 158-163.
[http://dx.doi.org/10.18474/0749-8004-44.2.158]

Ellis, J.D., Graham, J.R., Mortensen, A. (2013). Standard methods for wax moth research. *J. Apic. Res., 52*(1), 1-17.
[http://dx.doi.org/10.3896/IBRA.1.52.1.10]

El-Sawaf, SK (1950). The Life-history of the Greater Wax-moth (*Galleria mellonella* L.) In Egypt, with Special Reference to the Morphology of the mature Larva (Lepidoptera: Pyralidae). *Bulletin de la Societe Fouad 1er d'entomologie.*

Fasasi, K.A., Malaka, S.L. (2006). Life cycle and impact of greater wax moth, *Galleria mellonella* L.(Lepidoptera: Pyralidae) feeding on stored beeswax. *Niger J Entomol., 23*, 13-17.

Fedic, R., Zurovec, M., Sehnal, F. (2002). The silk of Lepidoptera. *Sanshi Konchuu Baiotekku, 71*(1), 1-5.

Ferguson, D. (1987). Lepidoptera. *Insect and mite pests in food: An illustrated key.* USDA Agriculture Handbook.

Finn, W.E., Payne, T.L. (1977). The attraction of greater wax moth females to male-produced pheromones. *Southwestern entomologist., 2*, 6.

Greenfield, M.D. (1981). Moth sex pheromones: An evolutionary perspective. *Fla. Entomol., 64*(1), 4-17.

[http://dx.doi.org/10.2307/3494597]

Hosamani, V., Hanumantha Swamy, B.C., Kattimani, K.N., Kalibavi, C.M. (2017). Studies on the biology of greater wax moth (*Galleria mellonella* L.). *Int. J. Curr. Microbiol. Appl. Sci., 6*(11), 3811-3815.
[http://dx.doi.org/10.20546/ijcmas.2017.611.447]

Jindra, M., Sehnal, F. (1989). Larval growth, food consumption, and utilization of dietary protein and energy in *Galleria mellonella. J. Insect Physiol., 35*(9), 719-724.
[http://dx.doi.org/10.1016/0022-1910(89)90091-7]

Klein, A.M., Vaissière, B.E., Cane, J.H., Steffan-Dewenter, I., Cunningham, S.A., Kremen, C., Tscharntke, T. (2007). Importance of pollinators in changing landscapes for world crops. *Proc. Biol. Sci., 274*(1608), 303-313.
[http://dx.doi.org/10.1098/rspb.2006.3721] [PMID: 17164193]

Kong, H.G., Kim, H.H., Chung, J., Jun, J., Lee, S., Kim, H.M., Jeon, S., Park, S.G., Bhak, J., Ryu, C.M. (2019). The *Galleria mellonella* hologenome supports the microbiota-independent metabolism of long-chain hydrocarbon beeswax. *Cell Rep., 26*(9), 2451-2464.e5.
[http://dx.doi.org/10.1016/j.celrep.2019.02.018] [PMID: 30811993]

Kumar, G., Khan, M.S. (2018). Study the life cycle of greater wax moth (*Galleria mellonella*) under storage conditions in relation to different weather conditions. *J. Entomol. Zool. Stud., 6*, 444-447.

Kwadha, C.A., Ong'amo, G.O., Ndegwa, P.N., Raina, S.K., Fombong, A.T. (2017). The biology and control of the greater wax moth, *Galleria mellonella. Insects, 8*(2), 61.
[http://dx.doi.org/10.3390/insects8020061] [PMID: 28598383]

Leyrer, R.L., Monroe, R.E. (1973). Isolation and identification of the scent of the moth, *Galleria mellonella*, and a revaluation of its sex pheromone. *J. Insect Physiol., 19*(11), 2267-2271.
[http://dx.doi.org/10.1016/0022-1910(73)90143-1]

Martel, A.C., Zeggane, S., Drajnudel, P., Faucon, J.P., Aubert, M. (2006). Tetracycline residues in honey after hive treatment. *Food Addit. Contam., 23*(3), 265-273.
[http://dx.doi.org/10.1080/02652030500469048] [PMID: 16517528]

Nielsen, R.A., Brister, C.D. (1979). Greater wax moth: Behavior of larvae. *Ann. Entomol. Soc. Am., 72*(6), 811-815.
[http://dx.doi.org/10.1093/aesa/72.6.811]

Oldroyd, B.P. (1999). Coevolution while you wait: Varroa jacobsoni, a new parasite of western honeybees. *Trends Ecol. Evol., 14*(8), 312-315.
[http://dx.doi.org/10.1016/S0169-5347(99)01613-4] [PMID: 10407428]

Oldroyd, B.P. (2007). What's killing American honey bees? *PLoS Biol., 5*(6), e168.
[http://dx.doi.org/10.1371/journal.pbio.0050168] [PMID: 17564497]

Opoosun, O.O., Odebiyi, J.A. (2009). Life Cycle Stages of Greater Wax Moth, *Galleria mellonella* (L.)(Lepidoptera: Pyralidae), Ibadan, Oyo State, Nigeria. *J. Entomol.., 26*, 21-27.

Paddock, F.B. (1918). *The beemoth or waxworm..* Texas Agricultural and Mechanical College System.
[http://dx.doi.org/10.5962/bhl.title.57185]

Pastagia, J.J., Patel, M.B. (2007). Biology of *Galleria mellonella* L. on brood comb of *Apis cerana* F. *J. Plant Prot. Environ., 4*(2), 85-88.

Rahman, A., Bharali, P., Borah, L., Bathari, M., Taye, R.R. (2017). Postembryonic development of *Galleria mellonella* L. and its management strategy. *J. Entomol. Zool. Stud., 5*(3), 1523-1526.

Roh, S.J., Park, H., Kim, S.H., Kim, S.Y., Choi, Y.S., Song, J.H. (2020). A new species of *Galleria* Fabricius (Lepidoptera, Pyralidae) from Korea based on molecular and morphological characters. *ZooKeys, 970*, 51-61.
[http://dx.doi.org/10.3897/zookeys.970.54960] [PMID: 33024409]

Shaik, H.A., Mishra, A. (2017). Silk recycling in larvae of the wax moth, *Galleria mellonella* (Lepidoptera:

Pyralidae). *Eur. J. Entomol.,* 114.

Sharma, V., Mattu, V.K., Thakur, M.S. (2011). Infestation of Achoria grisella F.(wax moth) in honeycombs of *Apis mellifera* L. Shiwalik Hills, Himachal Pradesh. *Int. J. Sci. Nat., 2*(2), 407-408.

Smith, T.L. (1965). External morphology of the larva, pupa, and adult of the wax moth, *Galleria mellonella* L. *J. Kans. Entomol. Soc.,* 287-310.

Spangler, H.G. (1987). Acoustically mediated pheromone release in *Galleria mellonella* (Lepidoptera: Pyralidae). *J. Insect Physiol., 33*(7), 465-468.
[http://dx.doi.org/10.1016/0022-1910(87)90109-0]

Spangler, H.G. (1986). Functional and temporal analysis of sound production in *Galleria mellonella* L. (Lepidoptera: Pyralidae). *J. Comp. Physiol. A Neuroethol. Sens. Neural Behav. Physiol., 159*(6), 751-756.
[http://dx.doi.org/10.1007/BF00603728]

Spangler, H.G. (1988). Sound and the moths that infest beehives. *Fla. Entomol., 71*(4), 467-477.
[http://dx.doi.org/10.2307/3495006]

Swamy, B.H. (2008). Bionomics and biometrics of Greater wax moth *Galleria mellonella* Linnaeus. *Asian J. Biol. Sci., 3*(1), 49-51.

Warren, L.O., Huddleston, P. (1962). Life history of the greater wax moth, *Galleria mellonella* L., in Arkansas. *J. Kans. Entomol. Soc., 35*(1), 212-216.

Williams, J.L. (1997). Insects: Lepidoptera (moths). In: Morse, R., Flottum, K., (Eds.), *Honey bee pests, predators, and diseases.* (pp. 121-141). Ohio, USA: The AI Root Company.

CHAPTER 3

Sequential Developmental Events in the Wax Moth Life Cycle

Abstract: The wax moth (WM) is a holometabolous insect with developmental stages of egg, larva, pupa, and adult in its life cycle. The development coherence of the wax moth is influenced by different abiotic and biotic environmental cues, including the larval diet, temperature, cannibalism, genomic content, insect hormones, and pheromones. The fecundity and fertility are comparatively high in WM to ensure species' survival within the honey bee hive. The wax moth adults preferentially infest the weaker colonies at night, where they live in the concealed space, usually on the top bar of the wooden chamber. Mating usually takes place on nearby trees; after that, the gravid female enters the hive to oviposit in the crevices and cracks to hide from the host honey bees. Afterward, the eggs hatch into the larva that feeds on the bee wax, honey, pollen, and exuviate of the honey bees. Severe localised concealing sites of WM in the hive facilitate specific pest protection in the host honey bee colony, eventually destroying the entire hive and forcing the honey bees to abscond the hive. The present chapter elucidates the development of a specific devastating pest of honey bee colonies, including influential abiotic and biotic factors. Furthermore, the differentiation of the life cycles of the greater wax moth (GWM) and the lesser wax moth (LWM) is also speculated in detail as per available literature.

Keywords: Words: Galleria mellonella, Honey Bee Colony, Wax Moth Development, Wax Moth Growth Influencing Factors.

INTRODUCTION

The composite colony of the honey bee is comprised of a polyandrous queen honey bee, thousands of worker honey bees, and a few hundred facultative residents, the drone honey bees. The eusocial insect constructs a well-organised hive with properly formed hexagonal cells for brood, honey, and storage. The created honey bee colony is damaged by two major pests, the lesser wax moth, *Achoria grisella* (*A. grisella*) and the greater wax moth, *Galleria mellonella* (*G. mellonella*), in the tropical and subtropical regions. These two prominent pest moths differ in morphology, development, reproduction, and life cycle (Ellis *et al.*, 2013). The concerning pest feeds on the bee wax, thereby damaging the hexagonal cells carrying brood, honey, and pollen grain and eventually forcing the honey bee colony to abscond (3a-d) (Ellis *et al.* 2013, Tsegaye *et al.* 2014). The hives of *Apis cerana* and *Apis dorsata* are preferentially infected by the wax moths due to the enormous bee hive size, which can ensure food availability to the voraciously feeding larvae of the pest (Swamy *et al.* 2009). The wax moth attacks weak honey bee colonies, with twice peak infestation preferences; mid-late spring and late summer to fall, when colonies become weak due to the stress of environmental constraints and due to honey robbing by other bee species. Generally, the honey bee colonies become weak due to food scarcity, climatic factors, pesticide spray, poor queen quality, disease infection and/or pest infestation. The wax moths act as opportunistic or secondary invaders of the honey bee colony (Figs. **3a - d**). The laboratory colonisation of the wax moth can be maintained on yeast, glucose, and honey, as well as wheat flour, corn flour, and milk powder (Akbar *et al.*, 2004; Singh *et al.*, 2014). Warren and Huddleston, 1962, tried to culture the wax moth under the controlled conditions of the laboratory. After emergence, adults were paired in jars with a width of 1.5 inches and 3 inches. The female lays eggs in multiple broths received in the oviposition chamber. The hatched larvae were transferred to a similar jar with a culture medium for the growth of the larvae. The larvae had been maintained on the honey and glycerine. Similarly, Vijaykumar *et al.*, 2019 tried corn meal, wheat flour, milk powder, yeast tablets, honey, and glycerine for wax moth laboratory maintenance.

Fig. (3a). A standard colony without an infestation of the wax moth. The worker honey bees can be visualised in a specific click while performing different duties. The intact hexagonal wax cells loaded with the unripe honey and the brood are also visible.

Fig. (3b). A brood chamber click with the wooden frames. On either side of the foundation wax sheet, worker honey bees have constructed the hexagonal cells, which the colony uses to store honey, pollen, and brood.

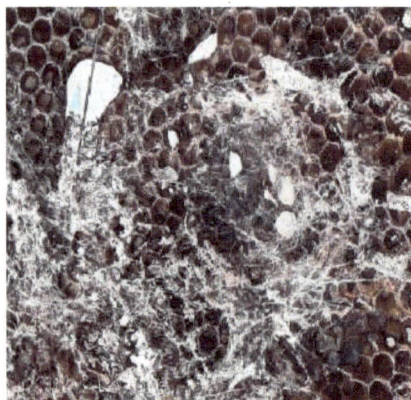

Fig. (3c). The wooden frames with the wax moth infestation. The silk threads are predominantly visible in the tunnels constructed by the wax moth larvae after feeding on the honey bee wax.

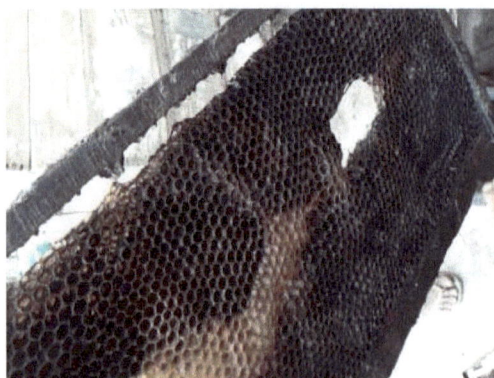

Fig. (3d). A wax moth-infested comb. The wax moth infestation force the host honey bees to abscond the hive and swarm.

The wax moth is a nocturnal insect that predominantly remains active at night. The female wax moths prefer to oviposit eggs in the cracks or the crevices. Therefore, there is little chance of laying eggs on the open surface (Milum, 1952). According to Milum and Geuther (1935), a wax moth female can lay eggs from 400 to 839, whereas Root (1950) reported that the wax moth female could lay about 1800 eggs with a depositing rate of 102 eggs per minute. The hatching of eggs and developmental conditions depend on environmental conditions, including temperature and humidity. According to Paddock, 1918, the eggs require approximately 7.2 to 21.8 days to hatch, with an average larval period of 35 to 110 days, while the pre-pupal phase requires about 5.75 to 6.4 days and the pupal period needs about 7.85 to 50 days. The life span of the female is about 12 days, whereas, for the male, it is 21 days in the greater wax moth (GWM).

Further, the same researchers reported that egg hatching was significantly affected at the temperature of 90°F and with 35% relative humidity. In contrast, Andrews, (1922) concluded that temperature at about 84 °F and 98.6 °F, larval development was normal and rapid. Higher temperatures affect the eggs' hatchability and the adult moth's life span. The temperature rise to 113 °F is lethal for the larvae and other developmental stages. Paddock (1918) also reported that temperatures below 32°F killed the larvae and adults, whereas larvae receiving cluster temperature survived the winter (Paddock, 1918; Hayes, 1936; Whitecomb and Warren, 1936). The elaborative description of the various developmental stages, as per scientific literature, is as follows

WAX MOTH EGG STAGE DEVELOPMENT

The LWM female lays ellipsoid, ovoid and pink-cream or white coloured eggs with wavy lines running diagonally at regular intervals whereas, in the greater wax moth, eggs are spherical, creamy-white (3e, f) (Paddock 1918; Arbogast *et al.* 1980; Ferguson, 1987; Williams, 1997; Swamy 2008; Sharma *et al.* 2011; Ellis, Graham and Mortensen 2013; Hosamani *et al.* 2017; Kwadha *et al.* 2017; Desai *et al.* 2019). The eggs of both wax moths, including the greater or lesser, vary in surface texture. Arbogast *et al.* (1980) conducted an SEM image-based study to compare the eggs of the lesser and greater wax moths.

Galleria mellonella females lay 50-150 eggs in a cluster (Williams, 1997; Ellis *et al.*, 2013; Kwadha *et al.*, 2017; Desai *et al.*, 2019). The low humidity prevents the hatchability of the eggs; on average, eggs require ten days to hatch (Warren and Huddleston, 1962; Paddock, 1918; Ellis *et al.*, 2013). However, the larva can be visualised as a dark ring within the egg before four days of hatching. Further, about 12 hours before hatching, a fully formed larva can be detected through the egg chorion (Paddock, 1918).

The eggs will not survive in extreme cold (at or below 0 °C for 4.5 hours) or in intense heat (at or above 46 °C for 70 minutes). The appropriate temperature range for developing the greater wax moth eggs is from 29°C-35°C. At lower temperatures, development becomes hampered. The development of the eggs gets significantly hampered at a temperature of 0°C or above 46 °C. At cold temperatures (18°C), egg development slows by about 30 days (Williams, 1997). According to Warren and Huddleston (1962), the wax moth egg stage lasts 10.2 days at a temperature of 80-85°F and a humidity of 25-35%. Furthermore, the egg stage lasts approximately 30.7 days at 92 °C and 51% RH.

In the GWM, eggs take 3-30 days to hatch, which depends on the environmental conditions, while in the LWM, about 7-22 days are required for egg stage completion, which is further dependent on the ecological conditions (Figs. **3e** and **f**).

Fig. (3e). The eggs of the wax moth on a section of bee comb.

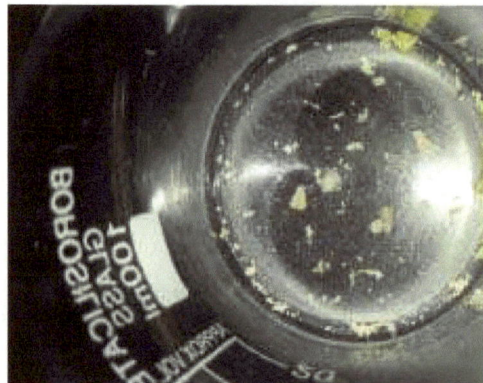

Fig. (3f). A click depicting the un-hatched eggs of the wax moth in a glass beaker in an attempt to colonise under laboratory conditions.

WAX MOTH LARVAL STAGE DEVELOPMENT

The duration of the larval development is variable in the GWM and the LWM, which is further dependent upon the temperature, food, other factors, and other optimal rearing conditions (Fig. **3g**). Furthermore, different studies found differences in developmental synchronicity based on environmental cues. For example, Chase (1921) reported that the wax moth larva passes through 8 or 9 instar stages at a temperature of 93 °F. On the other hand, several researchers reported a variable number of larval instars in G. mellonella, corresponding to a value of 5-10 (Swamy 2008; Hosamani *et al.* 2017; Desai *et al.* 2019). Moreover, according to Fasasi and Malaka (2006), there are only five larval instars. During development, the wax moth has seven larval instars, according to some researchers (Sehnal, 1966; Anderson and Mignat, 1970; Swamy, 2008; Ellis *et al.*, 2013; Venkatesh *et al.*, 2017; Desai *et al.*, 2019).

Fig. (3g). A click depicting the wax moth larvae and pupae enclosed within cocoons.

In the GWM, about 6-7 weeks at a temperature of 29° to 32°C are required for larval development, while the LWM completes larval development in 30.10 ± 2.5 days at a temperature of 29–33°C (Nielsen and Brister 1979; Williams 1997). The average duration required for the development of the different larval instars, including L1- L7, is 4.08, 5.72, 5.28, 6.96, 6.76, 7.64, and 8.40 days, respectively. About 45 days are required for larval development in the GWM (Desai *et al.*, 2019).

The larval body consists of the head, a three-segment thorax and an eleven-segmented abdomen (3g). The larva possesses a polipod body, with six thoracic and eight abdominal legs. The larval head carries short two-segmented antennae, chewing mouth parts, and four stemmata on either side (Ellis *et al.*, 2013). The retractable antennae are present, which are visible under the light microscope. The head capsule of the larva is yellow with a more significant prothoracic segment (Paddock, 1918). On the emergence of the larva, the thoracic legs are quite visible, whereas the abdominal legs become visible during 3rd day.

All these larval stages possess different body lengths, colours, growth rates, and food quantity requirements. The newly hatched larva starts feeding and spinning its web. The first larval instar is short, slim, and white, with a mean length of 1.27 mm (Hosamani *et al.* 2017). With further growth, it turns greyish-white. The third larval stage possesses a thickened body with massive and stocky ends (Arbogast *et al.*, 1980; Ferguson, 1987; Williams, 1997; Fasasi and Malaka, 2006; Sharma *et al.*, 2011; Ellis, Graham and Mortensen, 2013; Kwadha *et al.* 2017; Desai *et al.* 2019). The first instar larval body is weakly sclerotised without pigmentation except in the head region. In contrast, in later larval instars, the tergites of the pronotum, abdominal segments, protarsus, and claws become sclerotised with further development (Ellis *et al.*, 2013). In the final larval stage, a bright ecdysial line is visible on the mid-dorsal side (Kwadha *et al.*, 2017). The wax moth larvae are polypod and peripneustic, *i.e.,* nine pairs of spiracles.

The larva constructs an incision point in the cocoon for the escape of the adult after development. The greater wax moth larvae congregate in the hive, whereas the lesser wax moth is found individually within the comb (Paddock, 1918; Williams, 1997).

A chemical signal stimulates the larval moment and feeding. Paddock (1918) and Nielsen and Brister (1979) reported that the larvae isolated from the honeycombs returned to their food source. Cannibalism can occur in the absence of food in the case of the wax moth (Nielsen and Brister 1979; Williams 1997). The larvae of the wax moth feed on the bee wax, honey, exuviae of bee larvae, and pollen in the colony. Such food provides excellent energy and protein for proper growth (Kwadha *et al.*, 2017). If larvae obtain only a small amount of the protein, in that case, the process of silk spinning would be ceased due to the lack of the essential amino acids.

During the larval moulting, temporary feeding suspension usually occurs. During moulting, the old cuticle gets separated from the head capsule and the remaining larval body separately. As a result, the larvae reared on the artificial diet possess a comparatively longer larval period of development and a lower life span.

Furthermore, the protein content influences larval development. Mohamed *et al.* (2014) reported that larval growth is faster on old combs carrying pollen cells and larvae but slower on white or newly constructed honeycombs. According to Hosamani *et al.* (2017), foraging larval growth leads to the destruction of the honeycomb. However, the larvae can also be maintained on an artificial diet composed of milk powder, yeast, honey, and glycerol. Further, the diet's quality influences the larvae's immunity, as diet-deficient larvae are more susceptible to *Candida albicans Berhout* infections (Banville *et al.*, 2012). The greater wax moth larva constructs protective silken tubes, which are not detected by bees (Shaik *et al.*, 2017). These silken tunnels somewhere help in the protection of the wax moth larva. The larva moves to the cracks and crevices of the honeycomb to build protective silken tubes.

Consequently, the honeycomb gets destroyed. The silk composition used to construct the silken tunnels is similar to the silk used to build the cocoon. The core of the silk filament is composed of heavy and light chain fibroins and P25 chaperonin, with the filament coating formed of sericins.

WAX MOTH PRE-PUPAL DEVELOPMENT

The specific phase lasts 2 to 24 days, with an average time of 6.9 days (Paddorck, 1918). The female moth takes a comparatively longer time to pass through the pre-pupal period than the male moth.

WAX MOTH PUPAL DEVELOPMENTAL PERIOD

For the pupal phase, about 6-11 days are required. The female moth takes comparatively longer to complete the pupal stage than the male wax moth. Further, female spends an average of 8.5 days compared to male, with 7.9 days.

The larvae, after total growth, suspend feeding and search for a suitable place for the cocoon attachment and pupation. Commonly, the cocoon formation occurs within the wooden part of the hive. In a bee hive, the pupae can attach to the outer surface of the bee frames or to the inner lid of the hive, whereas in an abandoned hive, the cocoon gets attached anywhere within it (Paddock, 1918; Ellis *et al.*, 2013). About 2.5 days are required to construct the cocoon, which is further dependent upon the abiotic conditions of the environment (Paddock 1918). The cocoon acts as the protective enclosure for the pupa, which shields it from the worker bees, and various parasites and stabilises the abiotic conditions for further pupal development. The cocoon possesses a hard outer layer and a soft interior layer (Ellis *et al.*, 2013). The larva constructs an exit hole in the front portion of the cocoon for the exit of the adult. The larva closes the exit hole just before the pupation with a thin layer of silk (Paddock 1918; Desai *et al.* 2019).

Before the cocoon formation, the larva enters the pre-pupation phase by becoming slightly shrunken. The pre-pupation is not considered a separate stage like in other lepidopterans. The pre-pupa does not develop by moulting of the last larval instar.

Both phases, including the construction of the cocoon and the pupation event, are categorised into preparatory periods (Hosamani *et al.*, 2017; Desai *et al.*, 2019). During the pupation phase, imaginary discs' histogenesis occurs that comprises embryonic cells with rapid division potential. All these events are regulated by hormones (Chapman 1998). A few reports summarised the external morphology colour, sexual dimorphism, and dimension of *G. mellonella* pupa (Paddock 1918; Smith 1965; Swamy 2008; Hosamani *et al.* 2017; Kwadha *et al.* 2017; Desai *et al.* 2019).

The wax moth's pupa is obtect, meaning all appendages are cemented to the body through special secretion. During early pupation development, its colour is white, followed by yellow, brown and dark brown with further pupation. The pupa possesses a moderately elongated body with large eyes and long arched antennae that extend up to the second pair of wings (Smith 1965).

The ventral sides of abdominal segments VIII and IX are structurally different in female and male moths. The sclerite of segment VIII is separated in the female, whereas in the male, it is uniform. Further, in segment VIII, the female has a single copulatory aperture. In contrast, in the male, round knobs on this segment represent phallomeres and gonopore between them (Desai *et al.* 2019). The male pupa of the greater wax moth measures about 11.9–20 mm in length, and about 3.2–7 mm in width (Paddock 1918; Smith 1965; Swamy 2008; Ellis *et al.*, 2013; Hosamani *et al.* 2017; Kwadha *et al.* 2017; Desai *et al.* 2019), whereas the female pupa measures about length–15.83 and 11.86 mm; width–4.17 and 3.17 mm (Desai *et al.* 2019). The completion of the pupal stage is dependent upon the temperature and humidity, which can last for eight days at 28°C and 65% relative humidity, and up to 50 days at a temperature range from 2.5⁰C to 24⁰C, with a relative humidity of 44% to 100% (Swamy 2008; Hosamani *et al.* 2017; Kumar and Khan 2018; Desai *et al.*, 2019).

The developmental time of the greater wax moths from the larva to the pupa within the cocoon ranges from 3.75 days to 6.4 days, depending on the temperature (Figs. **3h** and **3i**). Inside the cocoon, the newly formed pupa is white and becomes yellow after 24 hours of pupation (Paddock, 1918). After four days, the pupa becomes a light brown that gradually becomes dark (Figs. **3g** and **h**). The greater wax moth pupa is 5 mm to 7 mm in diameter and 12 mm to 20 mm in length (Paddock, 1918). In the pupa, a row of spines develops from the back of the head to the fifth abdominal segment (Paddock, 1918; Williams, 1997).

Fig. (3h). The wax moth pupae in the enclosed form of a cocoon, and after taking it out of the cocoon.

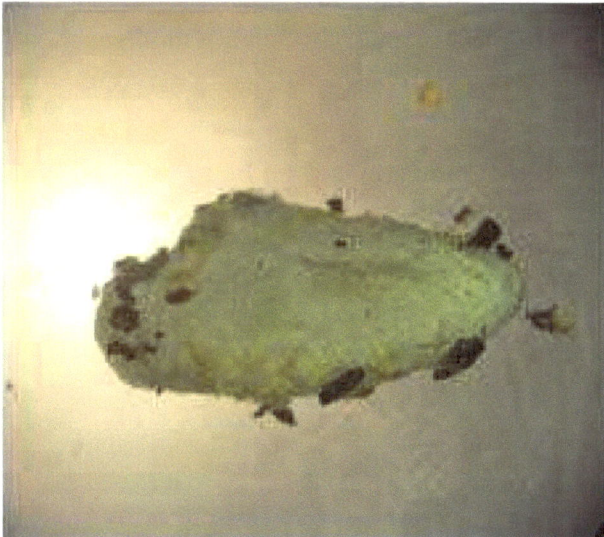

Fig. (3i). A cocoon of the wax moth covered with faecal matter for camouflage.

WAX MOTH ADULT DEVELOPMENT

During the emergence of an adult moth, it pushes out the silk lids of its cocoon for the exit (Swamy 2008). The newly hatched imago spends some inactive time extending and hardening its wings. The freshly emerged moth is white, which later on becomes grey (Smith, 1965; Leyrer and Monroe, 1973; Nielsen and Brister, 1979; Shimanuki, 1981; Williams, 1997; Swamy, 2008; Ellis *et al.*, 2013; Desai *et al.* 2019).

The adults are not capable of feeding due to the degenerated mouthparts; therefore, their life span is only 7-30 days in the ambient conditions (Figs. **3j** and **k**) (Paddock 1918; ElSawaf 1950; Opoosun and Odebiyi 2009; Hosamani *et al.* 2017; Kumar and Khan 2018). According to El-Sawaf (1950), the life span of the male wax moth is longer, with 21-30 days, than the female, with 8-15 days. Further, the female moth possesses three phases in her lifetime, including pre-oviposition, oviposition, and post-oviposition (Leyrer and Monroe 1973; Finn and Payne 1977; Nielsen and Brister, 1977; Greenfield 1981; Spangler, 1985, 1986; Spangler 1988; Jones *et al.* 2002; Desai *et al.* 2019), with a duration of 1.60 ± 0.50 days, 6.12 ± 1.09 days, and 2.00 ± 0.87 days, respectively (Desai *et al.* 2019). The mating in *G. mellonella* adults takes place in a different form.

Nielsen and Brister (1977) concluded that the copulation occurs on the nearby trees of the apiary; after that, the queen returns to the hives. According to Paddock (1918), after 24 hours of copulation, the female lays eggs at the concealed sites. According to Nielsen and Brister (1977), female moths usually enter the hive at night due to the less active host bees. During the evening, the entry of specific pests is discouraged by aggressive workers.

The wax moth female finds the crevices and cracks within the hive to lay eggs away from the consideration of the host. For ovipositing, the female moth extends its abdomen fully (Charriere and Imdorf 1999). By depositing eggs away from the reach of the bees, the wax moth female ensures the eggs' survival (Williams, 1997; Ellis *et al.*, 2013; Kwadha *et al.*, 2017). Hosamani *et al.* (2017) show that oviposition occurs between 19.00 and 03.00 h. Further, the wax moth's fertility depends upon the abiotic and biotic conditions (Mohamed *et al.* 2014).

A female moth lays about 50-180 eggs with a range of about 60 eggs per day (El-Sawaf 1950; Warren and Huddleston 1962; Hosamani *et al.* 2017), whereas in laboratory conditions, the total number of eggs ranges from 107 to 297 at 26.7°C, with 93.0% relative humidity as reported by Fasasi and Malaka 2006. Mohamed *et al.* (2014) observed that the fertility of the wax moth is related to the type of natural food and environmental conditions (Mohamed *et al.*, 2014; Kumar and Khan, 2018). Various abiotic conditions influence the total time required for overall development, which ranges from 32 days (28°C, 65% RH) to 93 days (2.5–24°C, 44–100% RH, food shortage) (Kumar and Khan 2018). The wax moth can produce 4-6 generations per year (Kwadha *et al.*, 2017). Longevity depends on environmental conditions like temperature and food (Arbogast *et al.*, 1980; Ferguson, 1987; Williams, 1997; Sharma *et al.*, 2011; Mohamed *et al.*, 2014; Kumar and Khan, 2018).

Fig. (3j). The adult of the greater wax moth with fully extended wings.

Fig. (3k). An adult of the lesser wax moth with extended wings.

LIFE CYCLE OF WAX MOTH

The wax moth being a holometabolous insect, includes the major developmental stages of egg, larva, pupa, and adult. The required time for conversion from egg to adult varies from a few weeks to a few months depending upon abiotic factors. Further, the development is affected by larval food, developmental duration, and cannibalism (Nielsen and Brister, 1979; Charriere and Imdorf, 1999; Gulati and Kaushik, 2004).

The average temperature range is 29–33 ^{0}C, and relative humidity is 29–33% for the development of the wax moth (Kwadha and Fombong, 2017).

Mating occurs on nearby trees, and the female returns to the hive with spermatophores (Nielsen and Brister, 1977). For mating, male moths release pheromones and acoustic sounds from tympanal organs to stimulate females, which respond by fanning the wings (Spangler, 1988).

Coskun *et al.* (2006) reported wax moth larvae could tolerate the different rearing diets without seriously affecting development. They tried to feed the wax moth larvae on a honeycomb and sucrose diet. The greater wax moth larvae develop within 40 days, with variability per environmental conditions. About 8-10 days are required for an egg to hatch, which is further affected by temperature, moisture, and nutrient quality (Ozer, 1962). Each female lays about 500 eggs throughout her life span. The egg hatches into the larva, and the mature larva can be 2.2 cm long. After the larval phase, they form pupa in the hive (Balazs, 1958; Beck, 1960; Bronksill, 1961; Dutky *et al.*, 1962; Townes *et al.*, 1965; Cantwell and Smith, 1970; Ali *et al.*, 1973; Burges, 1977, 1978; Sandlan, 1979a,b; 1980; 1982; Waage and Ng, 1984; Tsiropoulos, 1992; Başhan & Balcı, 1994; Kazmer & Luck, 1995; Ueno, 1999; King *et al.*, 1979; Taylor, 1988; Bratti & Costantini, 1991; Tsiropoulos, 1992; Gross, 1994; Gross *et al.*, 1996; Ueno, 1998; Harvey & Gols, 1998; Ueno, 1999; Nurullahoglu and Susurluk, 2001; King, 2002; Coskun *et al.*, 2005) . Furthermore, Gross *et al.* 1996 and King, 1987 reported that adding wheat flour and yeast to the wax moth larva can increase its weight.

The wax moth larva construct silk-lined tunnels throughout the hive, which destroy hexagonal wax cells, resulting in leakage of stored honey and damaging the brood and stored products. In addition, the larvae of the wax moth secrete silken thread, which creates hindrance in the moment of bees and entangles newly emerged bees that can die due to starvation. This phenomenon is described as *gallariasis*.

CONCLUSION

The wax moth exhibits complete metamorphosis with developmental stages including egg, larvae, pupae, and adult. The mating occurs outside the hive; after that, gravid females oviposit on the unattended comb of the bee colony. The larval forms are the most destructive stages, which eat up bee comb, eventually resulting in the abandonment of the hive by the host. The developmental phases of the GWM and the LWM can be defined by the variable developmental durations of the egg, larva, pupa, and adult stages. Further, developmental synchronicity is influenced by temperature, diet, and relative humidity by genomics, proteomics, and other environmental cues.

REFERENCES

Akbar, M., Nisa, Z.U., Janjua, A.S. (2004). Effect of some diets on longevity and fecundity of lesser wax moth. *Pak. J. Life Soc. Sci., 2*, 95-96.

Ali, A.D., Bakry, N.M., Abdellatif, M.A., El-Sawaf, S.K. (1973). The control of greater wax moth, *Galleria mellonella* L., by chemicals. *Z. Angew. Entomol., 74*(1-4), 170-177.
[http://dx.doi.org/10.1111/j.1439-0418.1973.tb01795.x]

Andrews, J.E. (1922). Some experiments with the larva of the wax moth *Galleria mellonella* L. *Trans. Wis. Acad. Sci. Arts Lett., 20*, 255-261.

Angelini, D.R., Kaufman, T.C. (2005). Insect appendages and comparative ontogenetics. *Dev. Biol., 286*(1), 57-77.
[http://dx.doi.org/10.1016/j.ydbio.2005.07.006] [PMID: 16112665]

Arbogast, R.T., Leonard Lecato, G., Van Byrd, R. (1980). External morphology of some eggs of stored-product moths (Lepidoptera pyralidae, gelechiidae, tineidae). *Int. J. Insect Morphol. Embryol., 9*(3), 165-177.
[http://dx.doi.org/10.1016/0020-7322(80)90013-6]

Balázs, A. (1958). Nutritional and nervous factors in adapting *Galleria mellonella* to artificial diet. *Acta Biol. Hung., 9*, 47-69.

Banville, N., Browne, N., Kavanagh, K. (2012). Effect of nutrient deprivation on the susceptibility of *Galleria mellonella* larvae to infection. *Virulence, 3*(6), 497-503.
[http://dx.doi.org/10.4161/viru.21972] [PMID: 23076277]

Başhan, M., Balci, K. (1994). The effects of wheat germ oil, vitamin E and some fatty acids on the reproduction of Melanogryllus desertus Pall. *Turk. J. Zool., 18*, 147-151.

Beck, S.D. (1960). Growth and development of the greater wax moth, *Galleria mellonella* (L.).(Lepidoptera: Galleriidae). *Trans. Wis. Acad. Sci. Arts Lett., 49*, 137-148.

Bratti, A., Costantini, W. (1991). Effects of new artificial host diets on the host-parasitoid system *Galleria mellonella* L.(Lepidoptera: Galleriidae) Archytas marmoratus Town.(Diptera: Tachinidae). *Redia (Firenze), 74*, 445-448.

Bronskill, J. (1961). A cage to simplify the rearing of the greater wax moth, *Galleria mellonella* (Pyralidae). *J. Lepid. Soc., 15*(2), 102-104.

Burges, H.D. (1977). Control of the wax moth *Galleria mellonella* on bee comb by H-stereotype V *Bacillus thuringiensis* and effect of chemical additives. *Apidologie (Celle), 8*(2), 155-168.
[http://dx.doi.org/10.1051/apido:19770206]

Burges, H.D. (1978). Control of wax moths: physical, chemical and biological methods. *Bee World, 59*(4), 129-138.
[http://dx.doi.org/10.1080/0005772X.1978.11097713]

Cantwell, G.E., Smith, L.J. (1970). Control the greater wax moth, *Galleria mellonella*, in honeycomb and comb honey. *Am. Bee J., 110*(4)

Charriere, J.D., Imdorf, A. (1999). Protection of honeycombs from wax moth damage. *Am. Bee J., 139*, 627-630.

Chase, R.W. (1921). The length of life of the larva of the wax moth, *Galleria mellonella* L., in its different stadia. *Trans. Wis. Acad. Sci. Arts Lett., 20*, 263-267.

Coskun, M., Kayis, T., Sulanc, M., Ozalp, P. (2006). Effects of different honeycomb and sucrose levels on the development of greater wax moth *Galleria mellonella* larvae. *Int. J. Agric. Biol., 8*(6), 855-858.

Coskun, M., Ozalp, P., Sulanc, M., Emre, I. (2005). Effects of Various Diets on the Oviposition and Sex Ratio of *Pimpla turionellae* L. *RNA, 75*, 75-00.

Desai, A.V., Siddhapara, M.R., Patel, P.K., Prajapati, A.P. (2019). Biology of greater wax moth, *Galleria mellonella*. On artificial diet. *J. Exp. Zool. India,* 22(2), 1267-1272.

Dutky, S.R., Thompson, J.V., Cantwell, G.E. (1962). A technique for mass rearing the greater wax moth. *Proc. Entomol. Soc. Wash., 64*(1), 56-58.

Ellis, J.D., Graham, J.R., Mortensen, A. (2013). Standard methods for wax moth research. *J. Apic. Res., 52*(1), 1-17.
[http://dx.doi.org/10.3896/IBRA.1.52.1.10]

Fasasi, K.A., Malaka, S.L. (2006). Life cycle and impact of greater wax moth, *Galleria mellonella* L.(Lepidoptera: Pyralidae) feeding on stored beeswax. *Niger J Entomol., 23*, 13-17.

(1987).

Gross, H.R., Rogers, C.E., Carpenter, J.E. (1996). Development of Archytas marmoratus(Diptera: Tachinidae) Reared in *Galleria mellonella* Larvae (Lepidoptera: Pyralidae) Feeding on Selected Diets. *Biol. Control, 6*(2), 158-163.
[http://dx.doi.org/10.1006/bcon.1996.0020]

Gross, H.R. (1994). Mass propagation of Archytas marmoratus (Diptera: Tachinidae). *Environ. Entomol., 23*(1), 183-189.
[http://dx.doi.org/10.1093/ee/23.1.183]

Gulati, R., Kaushik, H.D. (2004). Enemies of honeybees and their management—A review. *Agric. Rev. (Karnal), 25*(3), 189-200.

Harvey, J.A., Gols, G.J.Z. (1998). The influence of host quality on progeny and sex allocation in the pupal ectoparasitoid, *Muscidifurax raptorellus* (Hymenoptera: Pteromalidae). *Bull. Entomol. Res., 88*(3), 299-304.
[http://dx.doi.org/10.1017/S0007485300025906]

Hayes, W.P. (1936). Structural differences between greater and lesser wax moth. *J. Econ. Entomol., 29*(6), 1055-1058.
[http://dx.doi.org/10.1093/jee/29.6.1055]

Hosamani, V., Hanumantha Swamy, B.C., Kattimani, K.N., Kalibavi, C.M. (2017). Studies on the biology of greater wax moth (*Galleria mellonella* L.). *Int. J. Curr. Microbiol. Appl. Sci., 6*(11), 3811-3815.
[http://dx.doi.org/10.20546/ijcmas.2017.611.447]

Kazmer, D.J., Luck, R.F. (1995). Field tests of the size-fitness hypothesis in the egg parasitoid Trichogramma pretiosum. *Ecology, 76*(2), 412-425.
[http://dx.doi.org/10.2307/1941200]

King, E.G., Hartley, G.G., Martin, D.F., Smith, J.W., Summers, T.E., Jackson, R.D. (1979). *Production of the Tachinid Lixophaga Diatraeae on its Natural Host, the Sugar Cane Borer and on its Un-natural Host, the Greater Wax Moth.*. US Department of Agriculture, Science and Education Administration, Advances in Agricultural Research Service Handbook No. 693.

Kwadha, C.A., Ong'amo, G.O., Ndegwa, P.N., Raina, S.K., Fombong, A.T. (2017). The biology and control of the greater wax moth, *Galleria mellonella*. *Insects, 8*(2), 61.
[http://dx.doi.org/10.3390/insects8020061] [PMID: 28598383]

Leyrer, R.L., Monroe, R.E. (1973). Isolation and identification of the scent of the moth, *Galleria mellonella*, and a revaluation of its sex pheromone. *J. Insect Physiol., 19*(11), 2267-2271.
[http://dx.doi.org/10.1016/0022-1910(73)90143-1]

Milum, V.G. (1952). Geuther. Observations on the greater wax moth. *J. Econ. Ent., 28*, 576-578.
[http://dx.doi.org/10.1093/jee/28.3.576]

Mohamed, H.F., El-Naggar, S.E., Elbarky, N.M., Ibrahim, A.A., Salama, M.S. (2014). The impact of each of the essential oils of marjoram and lemon grass in conjunction with gamma irradiation against the greater wax moth, *Galleria mellonella*. *IOSR J. Pharm. Biol. Sci., 9*, 92-106.
[http://dx.doi.org/10.9790/3008-095492106]

Nielsen, R.A., Brister, C.D. (1979). Greater wax moth: Behavior of larvae. *Ann. Entomol. Soc. Am., 72*(6),

811-815.
[http://dx.doi.org/10.1093/aesa/72.6.811]

Nielsen, R.A., Brister, D. (1977). The greater wax moth: Adult behaviour. *Ann. Entomol. Soc. Am., 70*(1), 101-103.
[http://dx.doi.org/10.1093/aesa/70.1.101]

Nurullahoglu, U.Z., Susurluk, A.I. (2001). Fecundity of Turkish and German strains of *Galleria mellonella* (L.)(Lepidoptera: Pyralidae) reared on two different diets. *SU Fen-Edebiyat Fakültesi Fen Dergisi., 18*, 39-44.

Özer, M. (1962). Arı kovanlarında önemli zarar yapan balmumu güvesi *Galleria mellonella* L'nin morfoloji, biyoloji ve yayılışı üzerine araştırmalar. *Bitki Koruma Bul., 2*, 12-26.

Paddock, F.B. (1981). *The beemoth or waxworm.* Texas Agri. Expt. Sta. Bull.

Pastagia, J.J., Patel, M.B. (2007). Biology of *Galleria mellonella* L. on brood comb of *Apis cerana* F. *J. Plant Prot. Environ., 4*(2), 85-88.

Root, A.I. (1910). *The ABC and XYZ of bee culture..* AI Root Company.

Sandlan, K. (1980). Host location by Coccygomimus turionellae (Hymenoptera: Ichneumonidae). *Entomol. Exp. Appl., 27*(3), 233-245.
[http://dx.doi.org/10.1111/j.1570-7458.1980.tb02970.x]

Sandlan, K. (1979). Sex ratio regulation in Coccygomimus turionella Linnaeus (Hymenoptera: Ichneumonidae) and its ecological implications. *Ecol. Entomol., 4*(4), 365-378.
[http://dx.doi.org/10.1111/j.1365-2311.1979.tb00596.x]

Sandlan, K.P. (1982). Host suitability and its effects on parasitoid biology in Coccygomimus turionellae (Hymenoptera: Ichneumonidae). *Ann. Entomol. Soc. Am., 75*(3), 217-221.
[http://dx.doi.org/10.1093/aesa/75.3.217]

Sandlan, K.P. (1979). Host-feeding and its effects on the physiology and behaviour of the ichneumonid parasitoid, Coccygomimus turionellae. *Physiol. Entomol., 4*(4), 383-392.
[http://dx.doi.org/10.1111/j.1365-3032.1979.tb00631.x]

Sharma, V., Mattu, V.K., Thakur, M.S. (2011). Infestation of Achoria grisella F.(wax moth) in honey combs of *Apis mellifera* L. Shiwalik Hills, Himachal Pradesh. *Int. J. Sci. Nat., 2*(2), 407-408.

Shimanuki, H. (1981). *Controlling the greater wax moth: a pest of honeycombs..* US Dept. of Agriculture, Science and Education Administration.

Singh, S.P., Swati, R., Singh, J. (2014). Effect of artificial diet composition on some biological parameters of greater wax moth, *Galleria mellonella* L. under laboratory conditions. *J. Adv. Stud. Agric. Biol. Environ. Sci., 1*, 243-246.

Smith, T.L. (1965). External morphology of the larva, pupa, and adult of the wax moth, *Galleria mellonella* L. *J. Kans. Entomol. Soc., 287-310.

Sohail, M., Aqueel, M.A., Dai, P., Ellis, J.D. (2021). The larvicidal and adulticidal effects of selected plant essential oil constituents on greater wax moths. *J. Econ. Entomol., 114*(1), 397-402.
[http://dx.doi.org/10.1093/jee/toaa249] [PMID: 33558901]

Spangler, H.G. (1988). Sound and the moths that infest beehives. *Fla. Entomol., 71*(4), 467-477.
[http://dx.doi.org/10.2307/3495006]

Swamy, B.C., Venkatesh, H., Nagaraja, M.V. (2009). Influence of different species of honey bee combs on the life stages and biological parameters of greater wax moth, *Galleria mellonella* L. *Karnataka J. Agric. Sci., 22*(3), 670-671.

Taylor, A.D. (1988). Host effects on larval competition in the gregarious parasitoid Bracon hebetor. *J. Anim. Ecol., 57*(1), 163-172.
[http://dx.doi.org/10.2307/4770]

Townes, H, Monoi, S, Townes, M. A catalogue and reclassification of the eastern Palearctic Ichneumonoidae. *Mem. Amer. Ent. Inst., 5*

Tsegaye, A., Wubie, A.J., Eshetu, A.B., Lemma, M. (2014). Evaluation of different non-chemical wax moth prevention methods in the backyards of rural beekeepers in the North West dry land areas of Ethiopia. *IOSR J. Agric. Vet. Sci., 7*(3), 29-36.
[http://dx.doi.org/10.9790/2380-07312936]

Tsiropoulos, G.J., Anderson, T.E., Leppla, N.C. (1992). *Feeding and dietary requirements of the tephritid fruit flies.*. Boulder, CO: Westview Press.

Ueno, T. (1999). Multiparasitism and host feeding by solitary parasitoid wasps (Hymenoptera: Ichneumonidae) based on the pay-off from parasitised hosts. *Ann. Entomol. Soc. Am., 92*(4), 601-608.
[http://dx.doi.org/10.1093/aesa/92.4.601]

Vijayakumar, K.T., Neethu, T., Shabarishkumar, S., Nayimabanu Taredahalli, M.K., Bhat, N.S., Kuberappa, G.C. (2019). Survey, biology and management of greater wax moth, *Galleria mellonella* L. in Southern Karnataka, India. *J. Entomol. Zool. Stud., 7*(4), 585-592.

Waage, J.K., Ming, N.S. (1984). The reproductive strategy of a parasitic wasp: I. optimal progeny and sex allocation in Trichogramma evanescens. *J. Anim. Ecol., 53*(2), 401-415.
[http://dx.doi.org/10.2307/4524]

Warren, L.O., Huddleston, P. (1962). Life history of the greater wax moth, *Galleria mellonella* L., in Arkansas. *J. Kans. Entomol. Soc., 35*(1), 212-216.

Whitcomb, Jr Warren. "wax moth and its control." (1936).

WILLIAMS, J L (1997). Insects: Lepidoptera (moths). In: Morse, R., Flottum, K., (Eds.), *Honey bee pests, predators, and diseases.* (pp. 121-141). Ohio, USA: The AI Root Company.

CHAPTER 4

The Wax Moth Pheromone, Moth Influence, and Associated Glands

Abstract: The wax moth male secretes various pheromones to attract the female for mating. The volatiles induce species-specific influence, which modulates the behaviour of other members of the same species. The primary pheromones include nonanal and undecanal, hexanal, heptanal, octanal, decanal, undecanol, and 6, 10, 14 - trimethylpentadecanon-2. The specific chemicals are secreted by a pair of glands on the forewings of the male moth in the greater and lesser wax moth. These volatiles are essential for the adult stage and plays a critical role in larval and pupal aggregation. The specific chapter elaborates on the chemical composition of the pheromones, their influence on the conspecific individuals, and their role in modulating the mating behaviour, in the case of the greater wax moth (GWM) and the lesser wax moth (LWM).

Keywords: The Wax Moth Pheromone, Wax Moth Behaviour Change, Wax Moth Glands, Galleria mellonella.

INTRODUCTION

The nocturnal moths primarily depend upon the pheromones to locate the conspecific individuals. In most Lepidoptera, the female produced sex-specific volatiles involved in sexual communication. In contrast, a few exceptions are also in existence, where the male has pheromones to attract female moth, as in the case of *Galleria mellonella* (*G. mellonela*) and *Achroia grisella* (*A. grisella*) (Lofstedt *et al.*, 2016). In addition, the male wax moth releases a blend of volatile chemicals with acoustic sound to attract females of the same species (Leyrer and Monroe, 1973; Spangler, 1984).

The male sex pheromones have been identified in the different moth species, which induce specific courtship behaviour in the other conspecific members of the same species (Birch *et al.* 1990; Phelan 1997). In *G. mellonella* and *A. grisella*, a sex-role reversal has been reported in which the male secretes the sex pheromones to attract the female from a long distance. In wax moths, the male releases pheromones from the forewing glands and the ultrasonic signal that attracts.

the female wax moths. The male wax moth secretes two short aldehydes, mainly nonanal and undecanal (Leyrer and Monroe 1973; Romel *et al.* 1992).

Generally, male and female adults of the GWM and the LWM are unwanted guests of honey bee colonies. Strong colonies are less vulnerable to the wax moth infestation than colonies weakened by pesticide exposure or infection/infestation of any disease/pest (Romel *et al.* 1992). After emergence, adult moths leave the colony searching for mates, and mating takes place on nearby trees. The gravid female returns to the colony for oviposition (Nielsen and Brister 1977).

PHEROMONAL COMPOSITION OF WAX MOTH

In the scientific literature, numerous reports are procurable that are attributed to the chemical composition of the male wax moth pheromones. The greater wax moth emits pheromones to attract conspecific females over long distances and also produces ultrasonic calls. Primary volatiles extract from the calling male comprises a mixture of two aldehydes, nonanal and undecanal, as predominant components in a ratio of (7: 3) (Leyrer and Monroe 1973) and 1: 3 (Romel *et al.* 1992). The ratio of these two volatiles is variable and dependent upon differential ecological conditions. Further, in different locations, the ratio of nonanal and undecanal has been detected as 100: 50, 100: 60, 100: 70, 100: 80, 100: 100, and 80: 100. Other minor components include hexanal, heptanal, octanal, decanal, undecanol and 6,10,14 -trimethylpentadecanon-2 (Lebedeva *et al.*, 2002). The male sex pheromones include major aldehyde, namely nonanal and undecanal in 1:1, along with a minor component, 5,11-dimethylpentacosane (Flint and Merkle, 1983; Svensson *et al.*, 2014). Svensson *et al.* (2014) identified a male-specific pheromone 5,11-dimethylpentacosane. Furthermore, they identified a synergistic behavioural effect of this compound compared to aldehydes. Specific exploration indicated that very few female moths responded to aldehyde blend or to 5,11-dimethylpentacosane separately, but the female moth reacted positively to the combination of the two. They further reported that female moths responded more strongly to the male extract than the combination of these pheromones.

Svensson *et al.* (2014) reported the electrophysiological response of female wax moths to these compounds and concluded that females become attracted to these chemicals. Further, they analyzed that there was a third pheromone with a long-chain hydrocarbon 5,11- dimethylpentacosane. They reported that when female wax moths were exposed to a combination of nonanal and undecanal, or 5,11-dimethylpentacosane separately, very few exhibited behavioural responses, but when female wax moths were exposed to a combination of all these chemicals, they exhibited upwind orientation. Furthermore, the female wax moth was less responsive to the pheromonal variety than the male extract.

The male of the greater wax moth produces sex pheromones from the glands located on the forewings (Barth, 1937, Roller *et al.*, 1968). Finn and Pyne 1977; Flint and Merkle 1983; Ponomarev *et al.* 1997 observed, under laboratory conditions, that the female wax moth response had been detected to be comparatively more active to actual males than synthetic pheromonal bait.

After mating, the gravid female oviposits in the unprotected comb, which is less visited by the host honey bees. The insects detect the volatiles through the olfactory system in the hair-like sensilla on the antennae. Specific structures possess soluble and globular odorant binding proteins that facilitate the transportation of the odorants with specific critical tendencies, which are further dependent on the chain length and functional groups (Zhou, 2010; Venthur and Zhou, 2018).

Karen *et al.*,1992 identified the presence of aldehydes, primary alcohols, and fatty acids, *e.g.,* nonane and undecane. Gland extract of the greater wax moth indicated the presence of 19.0% undecanal, 3.9% nonanal, 48.3% 1-undecanol, and 28.8% 1-nonanol.

The sex pheromones of the wax moths, including n-nonanal and nundecanal in a ratio of 7:3, are secreted by the wing glands of the male as a response to the low-frequency sound produced by the female wing beat (Roller *et al.*, 1968; Leyrer and Monroe, 1973; Spangler, 1986, 1987). These aldehydes become easily oxidized soon after their release into the air, reducing specific volatiles' functionality. Therefore, somewhere, the continuous release of the pheromones is required. However, comparatively more stable esters of these aldehydes can be used for population control.

The bee alarm pheromones, including isopentyl acetate, benzyl acetate, octyl acetate, and 2-heptanone, induce a strong electroantennogram response in the female wax moth, even at low concentrations. Furthermore, octyl acetate induces a more robust response in the wax moth than isopentyl acetate, benzyl acetate, and 2-heptanone. Yuan *et al.*, 2019 tested electroantennogram responses, attraction/ repletion, and preference/avoidance of the GWM to alarm pheromones of the *Apis cerana*.

DETECTION OF THE PHEROMONES IN THE WAX MOTH FEMALE

Generally, moths possess two different groups of binding proteins that include pheromonal binding proteins (PBPs) and general odorant binding proteins (GOBPs) (Vogt *et al.*, 1991; Krieger *et al.*, 1996; Zhou *et al.*, 2009; Yin *et al.*, 2012; Liu *et al.*, 2015). Further explorations indicated that odorant binding proteins (OBP) are present in the moth's abdomen, proboscis, legs, or wings, which means functional diversification of these proteins (Gu *et al.*, 2015; Khurho *et al.*, 2017; Zhang *et al.*, 2017). Lizana *et al.*, 2020 concluded that comparatively fewer OBPs in the greater wax moth indicated a response to host specialization, as the honey bee uses different pheromones for communication.

The greater and fewer wax moths are the unwanted guests of the honey bee colony, forming the example of non-conformists. The greater wax moth secretes sex pheromones to attract the female from a long distance (Roller *et al.* 1968, Leyrer and Monroe 1973, Finn and Payne 1977). The functionality of these pheromones is probably different from aphrodisiac pheromones from the other moths (Birch 1974; Nielsen and Brister 1977; Greenfield 1981).

LARVAL PHEROMONES

The larvae of the GWM are quite destructive and exhibit gregarious behaviour that further aggravates the condition. Kwadha *et al.*, 2019 hypothesized that the larvae detect other conspecific larvae with the involvement of certain specific chemical cues. They had implemented a dual choice olfactometer assay to detect larval odours that have been used for conspecific aggregation among the 3-5th instar and 8th instar larvae. It has been noticed that the 8th instar larvae become attracted to odours from newly spun cocoons. Odour analysis using Coupled Gas Chromatography-Mass Spectrometry (GC MS) revealed the presence of four compounds in the larval head space: nonanal, decanal, tridecane, and tetradecane. Further, they had concluded that only the decanal induces significant attraction that facilitates the induction of larval aggregation.

Nielsen and Brister (1979) also reported the aggregation behaviour in the GWM larvae. The aggregation in the greater wax moth larvae and pupae probably provides the benefit of the future mate and protection from the natural enemies up to some limit. Although, the clustering of larvae can make the presence of the wax moth quite apparent to the natural predators and parasitoids. Further, the wax moth larva release pheromones like nonanal and un-decanal components. Kwadha *et al.*, 2019, detected through Dual-Choice Olfactometer Assays that 8th instar larvae got attracted to the cocoon-spinning larvae and to their head extract, indicating that specific chemical cues are responsible for the aggregation of larvae. The Third to fifth instar larvae possess more sensitivity to the odour from the food than the last instar larvae. Kwadha *et al.*, 2019 hypothesized that 3-5th instar larvae respond more to odours of the food, whereas the 8th instar larvae respond to the cocoon-spinning larval cues. The male moth sex pheromones include nonanal and decanal, heptanal, undeca- nal, and 6, 10, 14 trimethyl. Further, it had been detected that decanal acted as the predominant component of the larval aggregation pheromones, which escalates larval aggregation of 8th instar larvae (Kwadha *et al.*, 2019).

The GWM is considered a significant pest of bee hives (Jafari *et al.*, 2010; Nielsen and Brister, 1979). It drastically affects apicultural productivity by damaging the honey bee combs in tropical and sub-tropic regions (Burges, 1978; Williams, 1997). Neilsen and Brister (1979) observed that the wax moth adult, after emergence, just ran outside the hive and flew on trees or other places for mating during the first two days. After that, the gravid female returns to the hive for oviposition. The female of the GWM lays eggs for the first two days after mating. The wax moth remains active during the first three days, and with ageing, there is a reduction in the activities of the wax moth. Yuan *et al.* (2017) studied the physiological responses of GWM adults on the first, third, and fifth days after hatching. They concluded that with age, there is a reduction in the physiological activeness of wax moths. Wax moths use olfactory and auditory cues for mating, as research has shown that a lack of sound stimulation causes poor performance in females (Dahm *et al.*, 1971; Kindl *et al.*, 2011; Spangler, 1986, 1987; Svensson *et al.*, 2014). In addition, there is the implementation of semiochemicals for for intraspecific and interspecific communication (McNeil, 1991; Renou and Guerrero, 2000; Tak-;cs, Gries, and Gries, 2001; Cook *et al.*, 2007; Witzgall *et al.*, 2008). The greater wax moth secretes major undecanal (Roller *et al.*, 1968) and nonanal (Leyrer and Monroe, 1973), while minor components are decanal, hexanal, heptanal, 6,10,14 trimethylpen-; tacanol-;2 (Lebedeva *et al.*, 2002) and 5,11-;dimethylpentacosane (Svensson *et al.*, 2014). These male-specific pheromones attract virgin females without influencing the egg-laying potential of the female moth (Leyrer and Monroe, 1973; Flint and Merkle, 1983; Romel *et al.*, 1992; Lebedeva *et al.*, 2002).

CONCLUSION

A wide variety of pheromones are secreted by larval, pupal, and adult stages, influencing other members of the same species. The developmental stage-specific pheromones induce aggregation, whereas adult stage-specific pheromones induce the attraction of the female moth, mating, and reproduction. The primary pheromones include undecanal and nonanal, while minor components are decanal, hexanal, heptanal, 6,10,14 trimethylpen; tacanol-2, and 5,11-dimethylpentacosane This topic has received little attention in the case of wax moths; therefore, more research is needed to better understand the chemical communication, induced responses, and mechanisms involved in this insect.

REFERENCES

Barth, R. (1937). Structure and function of the wing glands of some microlepidoptera. Studies on the pyralides: *Aphomia gularis, Galleria mellonella, Plodia interpunctella*, Ephestia elutella und E. kühniella. *Z.*

Wiss. Zool., 150, 1-37.

Birch, M.C., Poppy, G.M., Baker, T.C. (1990). Scents and eversible scent structures of male moths. *Annu. Rev. Entomol., 35*(1), 25-54.
[http://dx.doi.org/10.1146/annurev.en.35.010190.000325]

Birch, M. (1974). Aphrodisiac pheromones in insects. In: Birch, M., (Ed.), *Pheromones* North-Holland Publications.(pp. 115-134). Amsterdam:

Burges, H.D. (1978). Control of wax moths: physical, chemical and biological methods. *Bee World, 59*(4), 129-138.
[http://dx.doi.org/10.1080/0005772X.1978.11097713]

Kwadha, C.A., Mutunga, J.M., Irungu, J., Ongamo, G., Ndegwa, P., Raina, S., Fombong, A.T. (2019). Decanal as a major component of larval aggregation pheromone of the greater wax moth, *Galleria mellonella. J. Appl. Entomol., 143*(4), 417-429.
[http://dx.doi.org/10.1111/jen.12617]

Cook, S.M., Rasmussen, H.B., Birkett, M.A., Murray, D.A., Pye, B.J., Watts, N.P., Williams, I.H. (2007). Behavioural and chemical ecology underlying the success of turnip rape (Brassica rapa) trap crops in protecting oilseed rape (Brassica napus) from the pollen beetle (Meligethes aeneus). *Arthropod-Plant Interact., 1*(1), 57-67.
[http://dx.doi.org/10.1007/s11829-007-9004-5]

Dahm, K.H., Meyer, D., Finn, W.E., Reinhold, V., Röller, H. (1971). The olfactory and auditory mediated sex attraction in *Achroia grisella* (Fabr.). *Naturwissenschaften, 58*(5), 265-266.
[http://dx.doi.org/10.1007/BF00602990] [PMID: 5580885]

Finn, W.E., Payne, T.L. (1977). The attraction of greater wax moth females to male-produced pheromones. *Southwestern entomologist.*

Flint, H.M., Merkle, J.R. (1983). Mating behaviour, sex pheromone responses, and radiation sterilization of the greater wax moth (Lepidoptera: Pyralidae). *J. Econ. Entomol., 76*(3), 467-472.
[http://dx.doi.org/10.1093/jee/76.3.467]

Greenfield, M.D., Karandinos, M.G. (1979). Resource partitioning of the sex communication channel in clearwing moths (Lepidoptera: Sesiidae) of Wisconsin. *Ecol. Monogr., 49*(4), 403-426.
[http://dx.doi.org/10.2307/1942470]

Gu, S.H., Zhou, J.J., Gao, S., Wang, D.H., Li, X.C., Guo, Y.Y., Zhang, Y.J. (2015). Identification and comparative expression analysis of odorant binding protein genes in the tobacco cutworm Spodoptera litura. *Sci. Rep., 5*(1), 13800.
[http://dx.doi.org/10.1038/srep13800] [PMID: 26346731]

Jafari, R., Goldasteh, S., Afrogheh, S. (2010). Control of the wax moth *Galleria mellonella* L. (Lepidoptera: Pyralidae) by the male sterile technique (MST). *Arch. Biol. Sci., 62*(2), 309-313.
[http://dx.doi.org/10.2298/ABS1002309J]

Khuhro, S.A., Liao, H., Zhu, G.H., Li, S.M., Ye, Z.F., Dong, S.L. (2017). Tissue distribution and functional characterization of odorant binding proteins in Chilo suppressalis (Lepidoptera: Pyralidae). *J. Asia Pac. Entomol., 20*(4), 1104-1111.
[http://dx.doi.org/10.1016/j.aspen.2017.07.010]

Kindl, J., Kalinová, B., Červenka, M., Jílek, M., Valterová, I. (2011). Male moth songs tempt females to accept mating: the role of acoustic and pheromonal communication in the reproductive behaviour of *Aphomia sociella.* PLoS One, 6*(10), e26476.
[http://dx.doi.org/10.1371/journal.pone.0026476] [PMID: 22065997]

Krieger, J., von Nickisch-Rosenegk, E., Mameli, M., Pelosi, P., Breer, H. (1996). Binding proteins from the antennae of Bombyx mori. *Insect Biochem. Mol. Biol., 26*(3), 297-307.
[http://dx.doi.org/10.1016/0965-1748(95)00096-8] [PMID: 8900598]

Lebedeva, K.V., Vendilo, N.V., Ponomarev, V.L., Pletnev, V.A., Mitroshin, D.B. (2002). Identification of pheromone of the greater wax moth *Galleria mellonella* from the different regions of Russia. *IOBC WPRS Bull., 25*(9), 229-232.

Leyrer, R.L., Monroe, R.E. (1973). Isolation and identification of the scent of the moth, *Galleria mellonella*, and a revaluation of its sex pheromone. *J. Insect Physiol., 19*(11), 2267-2271.
[http://dx.doi.org/10.1016/0022-1910(73)90143-1]

Liu, N.Y., Yang, K., Liu, Y., Xu, W., Anderson, A., Dong, S.L. (2015). Two general-odorant binding proteins in Spodoptera litura are differentially tuned to sex pheromones and plant odorants. *Comp. Biochem. Physiol. A Mol. Integr. Physiol., 180*, 23-31.
[http://dx.doi.org/10.1016/j.cbpa.2014.11.005] [PMID: 25460831]

Löfstedt, C, Wahlberg, N, Millar, JM (2016). Evolutionary patterns of pheromone diversity in Lepidoptera. *Pheromone communication in moths: evolution, behaviour and application, 1*, 43-82.

McNeil, J.N. (1991). Behavioural ecology of pheromone-mediated communication in moths and its importance in using pheromone traps. *Annu. Rev. Entomol., 36*(1), 407-430.
[http://dx.doi.org/10.1146/annurev.en.36.010191.002203]

Nielsen, R.A., Brister, D. (1977). The greater wax moth: Adult behaviour. *Ann. Entomol. Soc. Am., 70*(1), 101-103.
[http://dx.doi.org/10.1093/aesa/70.1.101]

Nielsen, R.A., Brister, C.D. (1979). Greater wax moth: Behavior of larvae. *Ann. Entomol. Soc. Am., 72*(6), 811-815.
[http://dx.doi.org/10.1093/aesa/72.6.811]

Lizana, P., Machuca, J., Larama, G., Quiroz, A., Mutis, A., Venthur, H. (2020). Mating-based regulation and ligand binding of an odorant-binding protein support the inverse sexual communication of the greater wax moth, *Galleria mellonella* (Lepidoptera: Pyralidae). *Insect Mol. Biol., 29*(3), 337-351.
[http://dx.doi.org/10.1111/imb.12638] [PMID: 32065441]

Phelan, PL (1997). Evolution of mate-signaling in moths: phylogenetic considerations and predictions from the asymmetric tracking hypothesis. *Evolution of mating systems in insects and arachnids.*
[http://dx.doi.org/10.1017/CBO9780511721946.015]

Ponomarev, V.L., Vendilo, N.V., Pletnev, V.A., Lebedeva, K.V., Melnikov, N.N. (1997). Influence of diet upon the composition of pheromone volatiles in *Galleria mellonella* (greater wax moth) males. *Arch. Insect Biochem. Physiol., 36*(2), 129-138.
[http://dx.doi.org/10.1002/(SICI)1520-6327(1997)36:2<129::AID-ARCH5>3.0.CO;2-Q]

Renou, M., Guerrero, A. (2000). Insect parapheromones in olfaction research and semiochemical-based pest control strategies. *Annu. Rev. Entomol., 45*(1), 605-630.
[http://dx.doi.org/10.1146/annurev.ento.45.1.605] [PMID: 10761591]

Roller, H., Biemann, K., Bjerke, J.S., Norgard, D.W., McShan, W.H. (1968). Sex pheromones of Pyralid moths. I. Isolation and identification of sex-attractant of *Galleria mellonella* L (greater wax moth). *Acta Entomol. Bohemoslov., 65*(3), 208-211.

Romel, K.E., Scott-Dupree, C.D., Carter, M.H. (1992). Qualitative and quantitative analyses of volatiles and pheromone gland extracts collected from *Galleria mellonella* (L.) (Lepidoptera: Pyralidae). *J. Chem. Ecol., 18*(7), 1255-1268.
[http://dx.doi.org/10.1007/BF00980078] [PMID: 24254163]

Spangler, H.G. (1986). Functional and temporal analysis of sound production in *Galleria mellonella* L. (Lepidoptera: Pyralidae). *J. Comp. Physiol. A Neuroethol. Sens. Neural Behav. Physiol., 159*(6), 751-756.
[http://dx.doi.org/10.1007/BF00603728]

Spangler, H.G. (1987). Acoustically mediated pheromone release in *Galleria mellonella* (Lepidoptera: Pyralidae). *J. Insect Physiol., 33*(7), 465-468.
[http://dx.doi.org/10.1016/0022-1910(87)90109-0]

Spangler, H.G. (1984). Responses of the greater wax moth, *Galleria mellonella* L.(Lepidoptera: Pyralidae) to continuous high-frequency sound. *J. Kans. Entomol. Soc.,* 44-49.

Svensson, G.P., Gündüz, E.A., Sjöberg, N., Hedenström, E., Lassance, J.M., Wang, H.L., Löfstedt, C., Anderbrant, O. (2014). Identification, synthesis, and behavioral activity of 5,11-dimethylpentacosane, a novel sex pheromone component of the greater wax moth, *Galleria mellonella* (L.). *J. Chem. Ecol.,* *40*(4), 387-395.
[http://dx.doi.org/10.1007/s10886-014-0410-8] [PMID: 24692052]

Takács, S., Gries, G., Gries, R. (2001). Communication ecology of webbing clothes moth: 4. Identification of male- and female-produced pheromones. *Chemoecology,* *11*(4), 153-159.
[http://dx.doi.org/10.1007/PL00001846]

Venthur, H., Machuca, J., Godoy, R., Palma-Millanao, R., Zhou, J.J., Larama, G., Bardehle, L., Quiroz, A., Ceballos, R., Mutis, A. (2019). Structural investigation of selective binding dynamics for the pheromone-binding protein 1 of the grapevine moth, *Lobesia botrana. Arch. Insect Biochem. Physiol.,* *101*(3), e21557.
[http://dx.doi.org/10.1002/arch.21557] [PMID: 31062883]

Vogt, R.G., Rybczynski, R., Lerner, M.R. (1991). Molecular cloning and sequencing of general odorant-binding proteins GOBP1 and GOBP2 from the tobacco hawk moth Manduca sexta: comparisons with other insect OBPs and their signal peptides. *J. Neurosci.,* *11*(10), 2972-2984.
[http://dx.doi.org/10.1523/JNEUROSCI.11-10-02972.1991] [PMID: 1719155]

Williams, JL (1997). Insects: Lepidoptera (moths). *Honey bee pests, predators, and diseases,* *3*, 119-142.

Witzgall, P., Stelinski, L., Gut, L., Thomson, D. (2008). Codling moth management and chemical ecology. *Annu. Rev. Entomol.,* *53*(1), 503-522.
[http://dx.doi.org/10.1146/annurev.ento.53.103106.093323] [PMID: 17877451]

Yang, S., Liu, W., Zhao, H., Du, Y., Pan, J., Wang, S., Guo, L., Xu, K., Jiang, Y. (2017). Observation on antennal sensilla of *Galleria mellonella* L. with scanning electron microscope. *Apic. China,* *68*, 16-19.

Yin, J., Feng, H., Sun, H., Xi, J., Cao, Y., Li, K. (2012). Functional analysis of general odorant binding protein 2 from the meadow moth, Loxostege sticticalis L. (Lepidoptera: Pyralidae). *PLoS One,* *7*(3), e33589.
[http://dx.doi.org/10.1371/journal.pone.0033589] [PMID: 22479417]

Li, Y., Jiang, X., Wang, Z., Zhang, J., Klett, K., Mehmood, S., Qu, Y., Tan, K. (2019). Losing the arms race: Greater wax moths sense but ignore bee alarm pheromones. *Insects,* *10*(3), 81.
[http://dx.doi.org/10.3390/insects10030081] [PMID: 30909564]

Zhang, Y.N., Zhu, X.Y., Ma, J.F., Dong, Z.P., Xu, J.W., Kang, K., Zhang, L.W. (2017). Molecular identification and expression patterns of odorant binding protein and chemosensory protein genes in *Athetis lepigone* (Lepidoptera: Noctuidae). *PeerJ,* *5*, e3157.
[http://dx.doi.org/10.7717/peerj.3157] [PMID: 28382236]

Zhou, J.J. (2010). Odorant-binding proteins in insects. *Vitam. Horm.,* *83*, 241-272.
[http://dx.doi.org/10.1016/S0083-6729(10)83010-9] [PMID: 20831949]

Zhou, J.J., Robertson, G., He, X., Dufour, S., Hooper, A.M., Pickett, J.A., Keep, N.H., Field, L.M. (2009). Characterisation of Bombyx mori Odorant-binding proteins reveals that a general odorant-binding protein discriminates between sex pheromone components. *J. Mol. Biol.,* *389*(3), 529-545.
[http://dx.doi.org/10.1016/j.jmb.2009.04.015] [PMID: 19371749]

CHAPTER 5

Mating and Reproduction in the Wax Moth: A Sequence of Events

Abstract: The wax moth adults prefer to mate on trees near the bee farm. For mating, they exhibit a specific behavioural pattern that includes the male's production of ultrasonic sound waves, wing vibration by the female moth, and pheromonal release by the male moth. After mating, the gravid female returns to the hive and oviposits there. The pairing in this moth occurs in a sex–role reversal manner rather than the typical moth signalling system. The male moth produces ultrasonic sound and pheromonal signals in this insect, whereas females have chemical signals in other moths. In other words, pairing in the wax moth occurs by releasing pheromones and wing fanning, attracting female moths with a response to wing fanning.

Furthermore, the presence of the female moth induces the male to produce ultrasonic sounds that attract the female and make her receptive to courtship. The current chapter elucidates signalling in the male wax moth, the response of the female exclusively to the male's mating calls and volatiles released by the male to guide her for the mating. Comparatively, more detailed information is available on the greater wax moth (GWM) than, the lesser wax moth (LWM) concerning insect biology, laboratory rearing, morphology, anatomy, physiology, genomics, proteomics, mating, reproduction, immunity, and plastic degradation capacity.

Keywords: Wax Moth Mating, Wax Moth Reproduction, Galleria mellonella.

INTRODUCTION

Galleria mellonella (*G. mellonella*), the greater wax moth (GWM), and *Achroia grisella* (*A. gresella*), the lesser wax moth, are considered ubiquitous troublesome challenges for apiculture due to the voracious feeding and destructive habits of various larval instars. The larvae of the concerned pest preferentially feed on the stored pollen grains and bee wax used to construct bee hives and exuviate the pupae and honey bee brood (Fig. **5a**). Furthermore, while feeding, the wax moth larvae construct silk-lined tunnels over the surface of the comb, and this larval habit creates a challenge for adult bees which get entangled within such a silken web. Within such a silken trap, the adult honey bees usually die due to starvation, a phenomenon known as *gallariasis*. An artificial wax moth population can be maintained in a laboratory under controlled conditions (Fig. **5b**). The wax moth larvae and adults are potential vectors of many diseases for honey bees. Severe infestation ultimately weakens the bee colony, which eventually absconds from the natural host of this pest (Kapil and Sihag, 1983 Hanumanthaswamy, 2000).

Fig. (5a). The pupae of the wax moth in a beaker under the controlled condition of the laboratory in BOD.

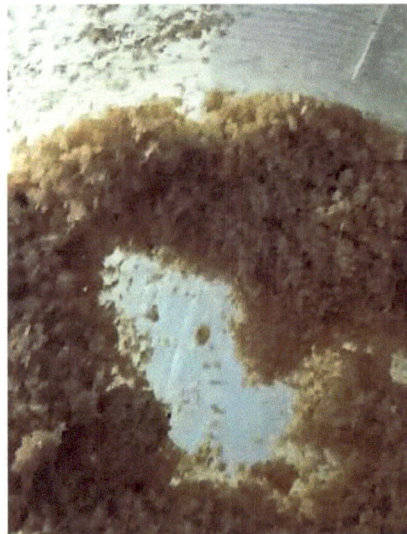

Fig. (5b). The wax moth culture under controlled laboratory conditions.

Galleria mellonella, the greater wax moth, is a notorious pest of the honey bee colonies with a worldwide distribution. Black-coloured excreta can be noticed in the infested colony as a significant indication of this insect. The substantial colony populations decline after severe infection by this pest, and destruction has been recorded (Hanumantha, 2000). In addition, a painful disease of the colony results in heavy economic losses for beekeepers (Kapil and Sihag, 1983; Hanumantha, 2007).

According to Greenfield and Coffelt (1983), after emergence, both the male and female adults move out of the hive entrance after eclosion. The mating occurs on the nearby trees, and the gravid female returns back to the hive for ovipositing. Further, they had reported that male moths, generally remain concealed beneath the hive, on low plants or on the trees to avoid desiccation. During the dusk, the male wax moths fly near the hives, possibly attracting females toward them. The gravid females lay eggs in the cracks of the fragile honey bee colonies. After eclosion, the adult male and female usually leave the colony but can stay near the colony if plenty of larval food is available.

Nielsen and Brister,1977 reported that during the scotophase, both freshly eclosed males and females flew to nearby trees, where they mate and, after that, gravid females return to honey bee colonies. The male moth produces acoustic sounds from tympanal organs, stimulating the female to fan her wings (Spangler, 1988). The female wax moth prefers males who advertise signals with higher amplitude and faster signal rate, *i.e.,* wingbeat frequency with longer silent gaps (Jang and Greenfield, 1996, 1998). The signalling males near each other usually modify their signals to attract the females (Eberhard, 1984).

In the wax moth, males produce pheromones and ultrasonic signals which attract female moths. Therefore pairing occurs in a 'sex-role reversal' of the typical moth signalling system (Greenfield and Coffelt, 1983; Spangler *et al.*, 1984; Spangler, 1985, 1987; Bennett *et al.*, 1991; Gwynne and Edwards, 1986; Heller and Krahe, 1994). In other moths, the pairing of male and female occurs by the searching male detecting the pheromone-emitting female, whereas in *Galleria mellonella* and *Achroia grisella*, the male produces acoustic signals that attract receptive females (Silberglied, 1977; Greenfield, 1981; Greenfield and Coffelt, 1983).

For mating, the male wax moth releases pheromones and ultrasound for 100–500 s, with a peak energy of 80 and 100 kHz, with emission bursts of 0.5-s at a repetition rate of 40 Hz with the involvement of tymbal organs (Spangler, 1985:1985, 1986, 1987). The mating calls of the male wax moth induce fanning in the female wax moth, which causes the male to secrete pheromones. Eventually, these sequential mating signals elicit the female to approach the male (Spangler, 1985, 1987). Jang and Greenfield, 1998, reported that signalling males usually form a cluster in high density and receptive female moths select males on the basis of their signal characteristics.

WAX MOTHS MATING AND THE INVOLVED EVENTS

The wax moth's specific mating behaviour includes the production of ultrasonic signals, wing fanning, pheromonal release, and mating (Fig. **5c**). The detail of mating is elaborated ahead.

Fig. (5c). An adult wax moth undergoing copulation in the laboratory under controlled conditions.

Ultrasonic Signals

Shortly after eclosion, the male moths of *G. mellonella* and *A. grisella* start doing wing fanning, which occurs during the scotophase phase in the first week of adulthood. After that, for the next four days, wing fanning activity reaches its peak after the onset of scotophase. Along with wing fanning, insects also exhibit antennal vibrating tendencies. The male wax moth occasionally shows antennal vibration during the scotophase without wing fanning. Peak activity of the wax moth has been recorded during the first half of scotophase, specifically 18:00–24:00 hours (Spangler, 1988). The male wax moth produces sound corresponding to the pressure level of 76 to 125 dB at a distance of 1cm from the hive. According to Jia *et al.* (2000), for mating *Achroia grisella*, the male produces ultrasonic signals which attract the female within 1-2 m. The male wax moths aggregate near the larval food resource and give indications for 6–10 h each night. The female selects the male based on signal rate (SR), loudness (peak amplitude; PA), and asynchrony interval (AI), which is a temporal feature. Further, they explored how the male wax moth modifies signals in the presence of the other signalers. According to Andersson (1994), the mature male competes with the other male for a mate.

G. mellonella and *A. grisella* produce ultrasonic pulses with the moment of their wings, which is caused by a pair of tymbals on the tegulae at the base of each forewing (Spangler *et al.*, 1984). With a slightly asynchronous wing moment, both tymbals produce two sound pulses during the upstroke and downstroke, separated by a brief gap of 100 to 1000 μs. At 25 C, sound vibrations have an amplitude of 95 dB (Spangler *et al.*, 1984; Snedden *et al.*, 1994); pulse pairs range from 70-140 s-1, with wing beat frequencies ranging from 35 to 70 s-1. Concomitantly with the wing fanning, the male wax moth secretes pheromones from glands on the forewings (Greenfield and Coffelt, 1983).

The observation that the male moths often begin a nightly signal only a few seconds, soon after their neighbours begin, indicates that they perceive and respond to their neighbours' signals. The male *A. grisella* significantly modifies the signal character. The modification probably occurs in the context of competition driven by female choice (Kunike, 1930). The female moth selects a male moth for mating based on mating signal characteristics.

Greenfield and Coffelt, 1982, demonstrated that the female wax moth even mated with a non-signalling male near the signalling males. If a male wax moth fails to match the other competitive male wax moths, it will lose the opportunity to mate with the female wax moth (Jang and Greenfield, 1998). The female greater wax moth responds by wing fanning to bursts of ultrasonic calls produced by the male wax moth (Skals and Surlykke, 2000).

Female Wax Moth's Discrimination between Ultrasonic Signals of the Wax Moth and the Bat

These ultrasonic calls of the wax moth exhibit some similarities with those of the echolocation calls of bats for food. Therefore, the female wax moth must be able to discriminate between these two sound frequencies.

The female wax moth exhibits a low-frequency response between 30 and 120 kHz, whereas, at 60 kHz, it exhibits comparatively greater sensitivity (Skals and Surlykke, 2000). In the male wax moth, the ultrasounds usually possess a 75 kHz-85 kHz sound frequency, whereas bat echolocation calls are of lower frequency (Spangler, 1986; Fenton *et al.*, 1998; Jones *et al.*, 2001).

Jones *et al.* (2002) studied the female wax moth's discrimination ability between sounds produced by the male wax moth and the bat. The wax moth produces ultrasonic signals for about 100–500s with a peak energy of between 80 and 100 kHz and an emission time of approximately 0.5-s bursts produced by tymbal organs (Spangler, 1985, 1986, 1987). The ultrasonic pulses are produced by the buckling in and snap out of the tymbal. With synchronous buckling, two pulses are produced per wing beat cycle, whereas four pulses can be produced with asynchronous wing beat.

It has been detected that the female wax moth responds by fanning her wings. Spangler (1985) reported wing fanning response in the female moth at 72 kHz synthetic ultrasound. In addition, the male calling induces fanning in the female, which further elicits pheromone release by the male.

Some bat species are susceptible to echoes of single-wing beats. Jia *et al.* (2000) reported that when ultrasonic sounds of the male wax moth and the bat echolocation calls were broadcast to the female wax moth, there was a reduction in display rate compared to display exhibited by female to male calls alone. It has been further reported that the male wax moth ceases calling in the presence of a bat echolocation call. Spangler, 1984 said that *Achroia grisella* finishes ultrasonic calling in the fact of 60 kHz ultrasonic pulse sound or when a bat flies over it. Greenfield and Weber (2000) state that the female wax moth differentiates bat echolocation calls from moth calls based on pulse rate.

Further, moth calls are shorter in duration than bat calls. The female prefers the male signals as she responds more to signals with longer periods and higher repetition rates (up to 71 Hz). Conclusions drawn from experiments conducted by Jia *et al.*, 2000 suggested that females can differentiate between the mating call of the wax moth and the predatory calls of bats.

Some of the bat species produce frequency-modulated call signals that the wax moth can easily detect, as Jones *et al.* (2002) reported that Daubenton's bat, *Myotis daubentonii*, emits sound in-between 90 and 30 kHz, with a similar pulse repetition rate as that of the male wax moth. The greater wax moth often displays mating signals outside the beehive; therefore possesses the risk of predation by gleaning bats. Further, they reported that in females, more wings fanning occurs when exposed to fewer echolocation bats calls. Therefore, the specified authors tried to induce wing fanning in the female moth by exposing them to the male calls concomitantly to varying bat calls with modulation of intensity. The mating calls of the male wax moth exhibit temporal and spectral similarities to those of the echolocation calls of bats. Therefore the female wax moth must be able to discriminate between mating and predatory calls. Jones *et al.*, 2002 analyzed the ability mentioned above of the female wax moth by exposing the female wax moth to mating calls of males, echolocation calls of the bat with variable intensity, and back calls of bats. The echolocation calls of bats and the mating calls of the male wax moth differ considerably in duration, which the moth nervous system can discriminate.

PHEROMONAL RELEASE

The wax moths exhibit non-conventional mating behaviour, where the male secretes pheromones to attract females. Kunike (1930) described mating behaviour in the wax moths as they release volatiles to attract females from a long distance. After that, Finn (1967) concluded that volatiles are released from a pair of glands located on the forewings. Dahm *et al.* (1971) identified that volatiles of the wax moth are composed of undecanal and cis-11-octadecenal, which are necessary for female attraction along with the acoustic signal produced by wing fanning in the males (Warren and Huddleston, 1962).

Male calls promote wing fanning by females, which in turn elicits pheromone production by males (Spangler, 1985, 1987). The wax moth male in high predation responds by ceasing pheromone release and exhibits reduced mating behaviour (Acharya and McNeil, 1998).

In different moths' sex, pheromone composition, tymbal moment, ultrasound frequency pattern, and intensity vary according to the species-specific pattern (Kindl *et al.*, 2011).

REPRODUCTION AND THE LIFE CYCLE

Galleria mellonella Linnaeus, the greater wax moth and *Achroia grisella* (*A. gresella*), the lesser wax moth, are ubiquitous pests of *Apis mellifera* Linnaeus, *Apis dorsata* and *Apis cerana* Fabricius. The wax moth is a cosmopolitan parasite that infests the weaker and older honey bee colonies. The mating occurs in the late evening or near the hive entrance. For pairing, the male produces pheromones and disperses them with the help of wing fanning, which produces sound (Greenfield and Coffelt, 1983). Snedden *et al.* (1994) analyzed that there is variation in the signal profiling of the wax moth from a population. The variation can be in the distribution of inter-pulse intervals, pulse duration, and frequency spectral peaks. The female probably considers the mating partner based on age, size, and signal characteristics (Snedden *et al.*, 1994).

The female wax moth lays eggs in clusters of 50–150 in cracks or crevices within the hive (Shimanuki, 1980; Shimanuki *et al.*, 1981; Akratanakul, 1987; Williams, 1997; Charriere and Imdorf, 1999; Ellis *et al.*, 2013). Further development of the wax moths occurs within the hive, enhancing their survival chances (Charriere and Imdorf, 1999). The eggs usually require 3–30 days to hatch into larvae (Shimanuki *et al.*, 1981; Williams, 1997; Charriere and Imdorf, 1999). The gravid females prefer weak colonies for egg laying. After egg hatching, the larva moves out from these hiding places to the honeybee comb (Williams, 1997). The larvae feed on honey, pollen, and brood, displaying aggregation and cannibalism (Nielsen and Brister, 1979; Williams, 1997). The early instars exhibit more extensive feeding than late instars (Nielsen and Brister, 1979). The larval phase is completed within 8–10 moulting stages, with the characterization of silk thread spinning, whereas the last instar spins the cocoon (Charriere and Imdorf, 1999). For pupation, larvae usually take 28 days to 6 months, depending upon the environmental conditions. The greater wax moth pupa remains enclosed within the cocoon and completes pupation in 1–9 weeks (Paddock, 1918; Williams, 1997; Ellis *et al.*, 2013). Within a year, the wax moth completes its 4-6 generations. Further, it can enter into reproductive diapause as an egg, larva, or pupa. (Kwadha, *et al.*, 2017).

The larvae get hatched in the morning between 08.30 and 11.00 h (Paddock, 1918: Oldroyd, 1999; 2007; Martel *et al.*, 2006; Klein *et al.*, 2007; Pastagia and Patel, 2007; Swamy, 2008; Hosamani *et al.* 2017; Desai *et al.* 2019; Kong *et al.* 2019). The voracious feeder larvae of the wax moth prefer to take pollen from bee brood and honey by burrowing through the edges of the unsealed cells of the honey bee comb. While moving through the comb, the larvae construct web tunnels, which trap the adult bees, resulting in *gallariasis*. The induced damage results in the absconding of colonies. Damage caused by the wax moth is more prevalent in tropical and subtropical regions.

At a temperature range of 29–32 °C, about five days are required for the egg to hatch, 6-7 weeks for larval phase completion and two weeks for pupal phase development, respectively. At 33.8 C, GWM goes through 8-9 moults (Charriere and Imdorf, 1999; Sohail *et al.*, 2017). According to Verma and Raj, 2000, the suitable temperature range is 16.6 ^{0}C-28.8 ^{0}C with a relative humidity of 43.2-81.6%, respectively. According to Jyothi (2003), 26 ± 2°C is the optimum temperature range for the growth and development of the specific insect. According to Hosamani *et al.*, 2017 seven successive larval instars develop in 4.50±0.49, 5.30±0.50, 6.60±0.68, 7.30±0.50, 8.30±0.45, 8.50±0.67, and 9.30±0.40 days respectively.

About 50.3±3.40 days are required in total for the larval phase development. For the pre-pupa and pupa phases, back 2.20 ± 0.53 and 8.65 ±0.73 days, respectively, are needed. The male adult life span is 16.50 ±2.70 days, whereas the female lives for 6.88 ±0.73 days. The male moth attracts the female with two different pheromones, concomitantly with short pulses of sound with a frequency of 75 kHz (Finn and Payne 1977; Greenfield 1981). The ultrasonic signal is generated with an acoustic signal, usually from structures found on the wings (Spangler, 1985, 1986). The female moth responds by fanning its wings, and the male responds by releasing pheromones, which attract the female toward them (Leyrer and Monroe, 1973; Spangler, 1985, 1986, 1987, 1988; Jones *et al.* 2002). The male moth produces sound impulses after sunset to attract the female. Spangler (1986) reported that sound waves are not produced in the presence of their natural hosts.

Additionally, the specific pest, 'the wax moth,' serves as a factitious host for several larval parasitoids and entomopathogenic nematodes, which can be reared on the wax moth (Ehlers and Shapiro-Ilan, 2005). In the laboratory, the wax moth can be easily cultured on an artificial diet under controlled conditions of insectary (5a,b) (Desai *et al.*, 2019)

CONCLUSION

The wax moth reproduces by a non-convenient method, as the male wax moth secretes volatile chemicals to attract the female wax moth. Usually, a single male mates with numerous females on the trees growing near the apiary. For mating, the male wax moth attracts the female wax moths with ultrasonic sound signals and pheromones. After mating, the females return to the hive, where they oviposit on unattended areas of bee combs. The hatching of eggs resulted in voracious feeding larvae that destroyed bee combs and stored pollen, honey, and the brood. Eventually, the destruction of the hive results in the abandonment of the colony.

REFERENCES

Acharya, L., McNeil, J.N. (1998). Predation risk and mating behavior: the responses of moths to bat-like ultrasound. *Behav. Ecol., 9*(6), 552-558.
[http://dx.doi.org/10.1093/beheco/9.6.552]

Akratanakul, P. (1987). *Honeybee diseases and enemies in Asia: a practical guide..* Food & Agriculture Org.

Alexander, D. (1975). Natural selection and specialized chorusing behaviour in acoustical insects. *Insect, science and society,* 35-77.

Andersson, MB (1994). Sexual selection: Princeton Univ Pr.

Bennett, A.L.A.P. (1991). On communication in the African sugarcane borer, Eldana saccharina Walker (Lepidoptera: Pyralidae). *J. Entomol. Soc. South. Afr., 54*(2), 243-259.

Charriere, J.D., Imdorf, A. (1999). Protection of honey combs from the wax moth damage. *Am. Bee J.*

Dahm, K.H., Meyer, D., Finn, W.E., Reinhold, V., Röller, H. (1971). The olfactory and auditory mediated sex attraction in *Achroia grisella* (Fabr.). *Naturwissenschaften, 58*(5), 265-266.
[http://dx.doi.org/10.1007/BF00602990] [PMID: 5580885]

Desai, A.V., Siddhapara, M.R., Patel, P.K., Prajapati, A.P. (2019). Biology of greater the wax moth, *Galleria mellonella*. On artificial diet. *J. Exp. Zool. India, 22*(2), 1267-1272.

Ehlers, R-U., Shapiro-Ilan, D.I. (2005). Mass production.*Nematodes as Biocontrol Agents.* (pp. 65-78). Wallingford, UK: CABI Publishing.
[http://dx.doi.org/10.1079/9780851990170.0065]

Ellis, J.D., Graham, J.R., Mortensen, A. (2013). Standard methods for wax moth research. *J. Apic. Res., 52*(1), 1-17.
[http://dx.doi.org/10.3896/IBRA.1.52.1.10]

Fenton, M.B., Portfors, C.V., Rautenbach, I.L., Waterman, J.M. (1998). Compromises: sound frequencies used in echolocation by aerial-feeding bats. *Can. J. Zool., 76*(6), 1174-1182.
[http://dx.doi.org/10.1139/z98-043]

Finn, W.E. (1967). *The Structure and Function of the Wing Gland in Achroia Grisella (Fabricius): Lesser*

The wax moth.. Madison: University of Wisconsin.

Greenfield, M.D., Coffelt, J.A. (1983). Reproductive behaviour of the lesser wax moth, *Achroia grisella* (Pyralidae: Galleriinae): signalling, pair formation, male interactions, and mate guarding. *Behaviour, 84*(3-4), 287-315.
[http://dx.doi.org/10.1163/156853983X00534]

Greenfield, M.D., Weber, T. (2000). Evolution of ultrasonic signalling in wax moths: discrimination of ultrasonic mating calls from bat echolocation signals and the exploitation of an antipredator receiver bias by sexual advertisement. *Ethol. Ecol. Evol., 12*(3), 259-279.
[http://dx.doi.org/10.1080/08927014.2000.9522800]

Greenfield, M.D. (1981). Moth sex pheromones: an evolutionary perspective. *Fla. Entomol., 64*(1), 4-17.
[http://dx.doi.org/10.2307/3494597]

Gwynne, D.T., Edwards, E.D. (1986). Ultrasound production by genital stridulation in Syntonarcha iriastis (Lepidoptera: Pyralidae): long-distance signalling by male moths? *Zool. J. Linn. Soc., 88*(4), 363-376.
[http://dx.doi.org/10.1111/j.1096-3642.1986.tb02253.x]

Heller, K.G., Krahe, R. (1994). Sound production and hearing in the pyralid moth Symmoracma minoralis. *J. Exp. Biol., 187*(1), 101-111.
[http://dx.doi.org/10.1242/jeb.187.1.101] [PMID: 9317421]

Jang, Y., Greenfield, M.D. (1998). Absolute versus relative measurements of sexual selection: assessing the contributions of ultrasonic signal characters to mate attraction in lesser the wax moths, *Achroia grisella* (Lepidoptera: Pyralidae). *Evolution, 52*(5), 1383-1393.
[http://dx.doi.org/10.1111/j.1558-5646.1998.tb02020.x] [PMID: 28565373]

Jang, Y., Greenfield, M.D. (1996). Ultrasonic communication and sexual selection in wax moths: female choice based on energy and asynchrony of male signals. *Anim. Behav., 51*(5), 1095-1106.
[http://dx.doi.org/10.1006/anbe.1996.0111]

Jia, F.Y., Greenfield, M.D., Collins, R.D. (2001). Ultrasonic signal competition between male the wax moths. *J. Insect Behav., 14*(1), 19-33.
[http://dx.doi.org/10.1023/A:1007893411662]

Jones, G., Barabas, A., Elliott, W., Parsons, S. (2002). Female greater wax moths reduce sexual display behavior in relation to the potential risk of predation by echolocating bats. *Behav. Ecol., 13*(3), 375-380.
[http://dx.doi.org/10.1093/beheco/13.3.375]

Kapil, R.P., Sihag, R.C. (1983). Waxmoth and its control. *Indian Bee J., 45*(2/3), 47-49.

Kindl, J., Kalinová, B., Červenka, M., Jílek, M., Valterová, I. (2011). Male moth songs tempt females to accept mating: the role of acoustic and pheromonal communication in the reproductive behaviour of *Aphomia sociella. PLoS One, 6*(10), e26476.
[http://dx.doi.org/10.1371/journal.pone.0026476] [PMID: 22065997]

Kunike, G. (1930). Zur biologie der kleinen wachsmotte, Achroea grisella Fabr. *Z. Angew. Entomol., 16*(2), 304-356.
[http://dx.doi.org/10.1111/j.1439-0418.1930.tb00139.x]

Nakano, R., Takanashi, T., Fujii, T., Skals, N., Surlykke, A., Ishikawa, Y. (2009). Moths are not silent, but whisper ultrasonic courtship songs. *J. Exp. Biol., 212*(24), 4072-4078.
[http://dx.doi.org/10.1242/jeb.032466] [PMID: 19946086]

Nielsen, R.A., Brister, D. (1977). The greater the wax moth: Adult behaviour. *Ann. Entomol. Soc. Am., 70*(1), 101-103.
[http://dx.doi.org/10.1093/aesa/70.1.101]

Paddock, F.B. (1918). *The beemoth or waxworm..* Texas Agricultural and Mechanical College System.
[http://dx.doi.org/10.5962/bhl.title.57185]

Shimanuki, H., Knox, D., Furgala, B., Caron, D., Williams, J. (1980). Diseases and pests of honey bees.

Beekeeping in the United States. *Agriculture Handbook., 335*, 118-128.

Shimanuki, H. (1981). *Controlling the greater the wax moth: a pest of honeycombs..* US Department of Agriculture, Science and Education Administration.

Silberglied, R.E. (1977). *How Animals Communicate.* (pp. 362-402). Bloomington: Indiana University Press.

Skals, N., Surlykke, A. (2000). Hearing and evasive behaviour in the greater wax moth, *Galleria mellonella* (Pyralidae). *Physiol. Entomol., 25*(4), 354-362.
[http://dx.doi.org/10.1046/j.1365-3032.2000.00204.x]

Snedden, W.A., Tosh, C.R., Ritchie, M.G. (1994). The ultrasonic mating signal of the male lesser wax moth. *Physiol. Entomol., 19*(4), 367-372.
[http://dx.doi.org/10.1111/j.1365-3032.1994.tb01065.x]

Spangler, H.G., Greenfield, M.D., Takessian, A. (1984). Ultrasonic mate calling in the lesser wax moth. *Physiol. Entomol., 9*(1), 87-95.
[http://dx.doi.org/10.1111/j.1365-3032.1984.tb00684.x]

Spangler, H.G. (1987). Acoustically mediated pheromone release in *Galleria mellonella* (Lepidoptera: Pyralidae). *J. Insect Physiol., 33*(7), 465-468.
[http://dx.doi.org/10.1016/0022-1910(87)90109-0]

Spangler, H.G. (1986). Functional and temporal analysis of sound production in *Galleria mellonella* L. (Lepidoptera: Pyralidae). *J. Comp. Physiol. A Neuroethol. Sens. Neural Behav. Physiol., 159*(6), 751-756.
[http://dx.doi.org/10.1007/BF00603728]

Spangler, H.G. (1988). Moth hearing, defence, and communication. *Annu. Rev. Entomol., 33*(1), 59-81.
[http://dx.doi.org/10.1146/annurev.en.33.010188.000423]

Spangler, H.G. (1984). Silence as a defence against predatory bats in two species of calling insects. *Southwest. Nat., 29*(4), 481-488.
[http://dx.doi.org/10.2307/3671001]

Spangler, H.G. (1985). Sound production and communication by the greater wax moth (Lepidoptera: Pyralidae). *Ann. Entomol. Soc. Am., 78*(1), 54-61.
[http://dx.doi.org/10.1093/aesa/78.1.54]

Swamy, B.H. (2008). Effect of colony strength and weather factors on the incidence of greater the wax moth (*Galleria mellonella* Linn.). *Asian Journal of BioScience., 3*(1), 81-83.

Warren, L.O., Huddleston, P. (1962). Life history of the greater the wax moth, *Galleria mellonella* L., in Arkansas. *J. Kans. Entomol. Soc., 35*(1), 212-216.

West-Eberhard, M.J. (1984). Sexual selection, competitive communication and species-specific signals in insects. *Proceedings of the 12th symposium of the Royal Entomological Society of London.*

Williams, J.L. (1997). Insects: Lepidoptera (moths). In: Morse, R., Flottum, K., (Eds.), *Honey Bee Pests, Predators, and Diseases* AI Root Company: Medina.(pp. 121-141). OH, USA:

CHAPTER 6

The Possible Wax Moth Infestation Regulation

Abstract: The wax moth infestation can be controlled with various physical, chemical, and biological methods. As the wax moth is a typical lepidopteran insect that is poikilothermic, exposing multiple developmental stages and imagoes to extremely low and higher temperatures can provide a solution for the specific pest infestation. Additionally, divergent pesticides targeting the nervous system, respiratory system, developmental regulation, and general insect physiology further solve the problem. The biological control measures include the application of various micro-organisms or their secreted products that can help regulate the insect population. The present chapter highlights primary methods for regulating the wax moth infestation with divergent strategies to reduce economic loss to apiculture.

Keywords: Physical Biological and Chemical Controls of the Wax Moth, Galleria mellonella, Pest control, Apiculture.

INTRODUCTION

The wax moth (Lepidoptera: Pyralidae), including Achoria grisella (the lesser wax moth (LWM)) and *Galleria mellonella* (the greater wax moth (GWM), are significant pests of bee hives, destroying bee wax inside and outside the colony (Paddock, 1918; Nielsen and Brister, 1977; Williams, 1997; Gillard, 2009; Ellis *et al.*, 2013; Tsegaye *et al.*, 2014). It is an opportunistic pest of weak colonies, where it invades and destroys hexagonal cells constructed for storage of brood, pollen grain, and honey (Dessalegn, 2001).

The stored combs provide a suitable habitat for the wax moth to breed. The wax moths are nocturnal; they become active at night, and during the day, the specific pest remains hidden in the hive (Gillard, 2009). The wax moth typically lives in areas of the hive that are dark, warm, poorly ventilated, and poorly defended by honey bees (Paddock, 1918; Williams, 1997; Ellis *et al.*, 2013). On entering the hive, the wax moth laid eggs on the unattended combs, and eggs take three days to hatch under normal conditions (Arbogast *et al.*, 1980; Chase, 1921; Eischen and Dietz, 1987; Smith, 1965; Warren and Huddleston, 1962). After hatching, larvae become bee wax comb pests and significantly damage the colony (Chase, 1921; Smith, 1965; Eischen and Dietz, 1987; Dessalegn, 2001). Different moth species are there, damaging wax combs in the storage. *Galleria mellonellu* L. (The greater wax moth) acts as a predominant pest that causes considerable damage to bee colonies. The wax moth causes significant impairment in colonies which become weak due to poor queen quality, exposure to insecticides for various reasons, or other causes. Further, the wax moth can damage combs that are improperly stored. In a colony, the wax moth lays eggs in crevices in the wooden body of the hive, which are not visited by honey bees frequently. The eggs hatch, and emerged larvae burrow into crevices of the comb by forming tunnels. The larvae destroy bee wax combs and create silken webs to spin combs together (Ali *et al.,* 1973). Furthermore, the spun silk web on the bee combs interferes with the normal activities of honey bees and brood development, resulting in colony absconding (Tsegaye *et al.,* 2014). The greater wax moth moults seven times throughout its story (Chase, 1921; Smith, 1965; Eischen and Dietz, 1987; Dessalegn, 2001).

There is an increase in the activity and the infestation rate of the wax moth during warmer climates (Crane, 2000). Therefore, the wax moth induces comparatively less damage in areas with temperatures below 25°C. However, the wax moth attacks the colonies, which become weak due to poor colony management (Dietemann *et al.,* 2013; Ellis *et al.,* 2013; Neumann *et al.,* 2013).

The wax moth infestation can be controlled by colony fumigation through a cotton cloth, tobacco leaf, and using different wax moth traps (Popolizio and Pailhe, 1973; Charriere and Imdorf, 1999; Shimanuki and Knox, 1997; Crane, 2000).

Abou-Shaara (2017) tried five different types of traps for the wax moth, which included a mesh envelope trap (MET), cup trap (CT), corrugated sheet (CS), wooden sheet trap (WST), and frame trap (FT). The wax moth can be treated by heat treatments, ozone treatment, freezing methods, climate manipulations, and chemical controls (Burges, 1978; Charrière and Imdorf, 1999; James, 2011; Ellis *et al.,* 2013). Other control measures include male sterilization with gamma rays, sex pheromone traps and light traps, substantial colonies maintenance, tobacco leaf smoke use, etc. (Jafari *et al.,* 2010; Sangramsinh *et al.,* 2014; Tsegaye *et al.,* 2014; Mabrouk and Mahbob 2015). Various other control strategies, such as the use of neem seed (Azadirachta indica A. Juss; Meliaceae), Beauveria bassiana, *Bacillus thuringiensis,* Berlin, and entomopathogenic nematodes, were found to be effective in some studies (El-Sinary and Rizk 2007; Ellis and Hayes 2009; Surendra *et al.,* 2010; Noosidum *et al.,* 2010; Abou-Shaara, 2017). Other methods include the use of red fire ants, *Bacillus thuringiensis,* cold treatment, male sterile technique (MST), fumigation, and others (Burges and Bailey, 1968a, b; Cantwell *et al.,* 1972; Vandenberg and Popolizio and Pailhe, 1973; Burgett and Tremblay, 1979; Goodman *et al.,* 1990; Shimanuki, 1990; Shimanuki and Knox, 1997;

Charriere and Imdorf, 1999; Crane, 2000; Ellis and Hayes, 2009; Hood *et al.*, 2003; Hood, 2010; Jafari *et al.,* 2010)

To avoid the wax moth attack to the colony, most effective measures include the sanitation, strong colony, adequate food sources, sealing of cracks and crevices, proper storage of wax, honey, and pollen stores; and additionally, beekeepers should avoid the use of pesticides near to the colony (Williams, 1997; Charriere and Imdorf, 1999; Gulati, 2004; Ritter, and Akratanakul, 2006).

PHYSICAL CONTROL

The physical methods include exposure of bee comb in the absence of bees to low and high temperatures. The exposure to low-temperature control method is an effective method when implemented safely.

The development of the wax moth can be prevented by exposing bee-keeping equipment and hive combs to different temperatures (Williams, 1997; Charriere and Imdorf, 1999; Gulati, 2004; Ritter and Akratanakul, 2006). Burges, 1978 suggested that controlling the wax moth infestation requires overnight exposure in deep domestic freezers, ten days at 2–30 C and three weeks at 5^0C. Further, the duration of the exposure is dependent upon the bulk of the material and the cooling capacity of the freezer. During stalking of infested frames, appropriate air space must be present for air circulation. For commercial bee-keeping, the infested comb can be exposed to a higher temperature of 45–80 °C for a time duration of 1-4 hours, whereas for small-scale farming, combs can be kept in hot water for 3-5 hours (Gulati, and Kaushik, 2004).

Further, the wax moth-infested combs can be exposed to lower temperatures −7 °C to −15 °C for 2–4.5 h (Gulati and Kaushik, 2004).

Similarly, high-temperature exposure can be used to kill infested combs and related structures (Cantwell and Lehnert, 1968; Cantwell and Smith, 1970; Cantwell and Mathenius, 1973). In addition, all developmental stages of the wax moth get killed by exposure to higher temperatures of up to 60 °C for 24 hours (Cantwell and Lehnert, 1968). Further, the wax moth can be controlled by giving exposure to gamma radiation, which hampers the normal reproduction of the wax moth (Burges, 1978).

CHEMICAL CONTROL

Various insecticides which are used to control the wax moth include allethrin, cidial, dimethoate, dioxathion, endosulfan, endosulfan-fenitrothion, ethion, Fac, imidan, imidan-endosulfan, pyrethrins, toxaphene, trichlorfon (Ali *et al.,* 1973).

Contact insecticides are preferred less because of lower penetration and harmful residue. The various fumigants used to control the wax moth include carbon disulfide, sulfur, calcium cyanide, methyl bromide, ethyl dibromide, ethylene oxide, par2 dichlorobenzene, hydrogen cyanide, and phosphine (Krebs, 1957; Ali *et al.,* 1973; Burges, 1978). Fumigant use is comparatively effective, as volatiles can enter different crevices throughout the comb.

Less preference is given to carbon disulfide and hydrogen cyanide due to their inflammability and extremely poisonous nature (Burges, 1978). Exposure of fumigants should be delivered in an air-tight structure so that volatiles can penetrate throughout the comb and/or equipment and/or structures. Dosage is determined by considering volatile gas concentration and exposure time. Therefore, exposure methods can include low concentration/long duration or high concentration/short duration depending upon toxicity, chemical nature, and effectiveness of selected chemicals (Burges, 1978). Further, the former method involves less gas usage, whereas the latter method involves less time for loss by leakage.

Additionally, leakage of fumigant can be prevented by sealing windows, doors, and other sites with adhesive tape and/or polythene sheets. The various volatiles is absorbed by the fumigated material and then released slowly. Therefore, exposed material should be kept in place with proper ventilation (Burges, 1978).

Various fumigants which are used for control of the wax moth include sulphur, acetic acid, ethylene bromide, calcium cyanide, formic acid methyl bromide, phosphine, paradichlorobenzene (PDB) naphthalene, and carbon dioxide (Charriere and Imdorf, 1999; Gulati and Kaushik, 2004; Ritter, and Akratanakul, 2006). The carbon dioxide can be used for the regulation of the wax moth. Further, fumigants can be applied to infested combs under air-tight conditions. These fumigants have fewer consequences than other control measures, but they pose a risk to handlers and leave residues in the environment, so they are avoided in many countries (Ritter and Akratanakul, 2006; Kwadha *et al.,* 2017).

Essential oils are preferred over synthetic insecticides to control the wax moth. These essential oils are complex mixtures of different phytochemicals that possess synergistic and antagonistic interactions among themselves (Ferreira *et al.,* 2017).

BIOLOGICAL CONTROL

Various biological controls include *Bacillus thuringiensis Berliner* (H-serotype V) (Bt), *Bracon hebetor* (Say), *Trichogramma* species, the red imported fire ant (RIFA) (*Solenopsis invicta Buren* and *Solenopsis germinita Fabricius*), and the use of the male sterile technique (MST). The lesser wax moth is comparatively resistant to *Bacillus thuringiensis* (Millar, 1965; Burges and Bailey, 1968; Burges, 1978). Dougherty (1982) assessed the efficacy of the nuclear polyhedrosis virus in controlling wax moth infestations and the cost of the specific control.

Biologically, the wax moth has been controlled by using *Bacillus thuringiensis* since the 1960s (Heitor, 1960; Euverte and Martouret, 1963; Burges and Bailey, 1968). The specific biological control is available in the form of powders and stabilized suspensions of the bacterium and bipyramidal crystals of toxic protein secreted by spores. The moth larvae consume the bacterial spores, and their germination causes lethal infection. These specific materials act as stomach poisons for the wax moth but are harmless for bees and humans (Johansen, 1962; Burges and Bailey, 1968; Johansen, 1968; Heimpel, 1971). The water suspension of these materials can be used to impregnate the foundation wax sheet during milling. Usually, the weight of spores and crystals should be about 1% less than the weight of bee wax in the comb (Burges, 1977). When used, such an impregnated foundation sheet results in leaching out into the first crop of honey, which does not affect honey bees or humans.

Burges (1977) concluded that Bt application is practical for the first season, but the wax moth larvae become less susceptible during the second and third seasons. Gulati and Kaushik (2004) reported that Bt was effective for 13 months only. However, resistance to Bt toxins has been reported in the Pyralidae family (Shelton *et al.,* 1993; Tabashnik *et al.,* 1994; McGaughey and Johnson, 2017; Kwadha *et al.,* 2017).

STERILIZATION

Jafari *et al.,* (2010) concluded that the most effective male sterilization technique was when the wax moth pupae were sterilized by using 350 Gy of gamma-radiation.

Some other control methods for the wax moths have been developed, especially for stored combs. These methods include using paradichlorobenzene crystals (Burges 1978; Charrière & Imdorf 1999). Recently, non-chemical traps have been developed to reduce damage due to the wax moths inside beehives (Abou-Shaara 2017). These include sex pheromone traps (Sangramsinh *et al.,* 2014) and light traps (Mabrouk & Mahbob 2015).

CONCLUSION

The wax moth infection can be regulated with divergent regulations, including mechanical removal, proper colony management, maintaining the queen's fecundity and fertility, strong colony, adequate storage for bee wax, and involvement of different chemicals and biological controls. Like other pest management strategies, integrated control comprising different control techniques can be used, which helps reduce the economic loss caused by the specific pest.

REFERENCES

Abou-Shaara, H.F. (2017). Evaluation of Non-Chemical Traps for Management of Wax Moth Populations within Honey Bee Colonies. *J. Agric. Urban Entomol., 33*(1), 1-9.
[http://dx.doi.org/10.3954/1523-5475-33.1.1]

Abou-Shaara, H.F. (2017). Greater Wax Moth Larvae Can Complete Development on Paper Wasp Nest. *J. Agric. Urban Entomol., 33*(1), 57-60.
[http://dx.doi.org/10.3954/JAUE17-13.1]

Meresa Lemma Evaluation of different non-chemical the wax moth prevention methods in the backyards of rural beekeepers in the North West dry land areas of Ethiopia. *IOSR Journal of Agriculture and Veterinary Science (IOSR-JAVS), 7*(3), 29-36.

Ali, A.D., Bakry, N.M., Abdellatif, M.A., El-Sawaf, S.K. (1973). The control of greater wax moth, *Galleria mellonella* L., by chemicals. *Z. Angew. Entomol., 74*(1-4), 170-177.
[http://dx.doi.org/10.1111/j.1439-0418.1973.tb01795.x]

Arbogast, R.T., Leonard Lecato, G., Van Byrd, R. (1980). External morphology of some eggs of stored-product moths (Lepidoptera pyralidae, gelechiidae, tineidae). *Int. J. Insect Morphol. Embryol., 9*(3), 165-177.
[http://dx.doi.org/10.1016/0020-7322(80)90013-6]

Burges, H.D., Bailey, L. (1968). Control of the greater and lesser wax moths (*Galleria mellonella* and *Achroia grisella*) with *Bacillus thuringiensis*. *J. Invertebr. Pathol., 11*(2), 184-195.
[http://dx.doi.org/10.1016/0022-2011(68)90148-1] [PMID: 5672009]

Burges, H.D., Thomson, E.M., Latchford, R.A. (1976). Importance of spores and δ-endotoxin protein crystals of *Bacillus thuringiensis* in *Galleria mellonella*. *J. Invertebr. Pathol., 27*(1), 87-94.
[http://dx.doi.org/10.1016/0022-2011(76)90032-X] [PMID: 932482]

Burges, H.D. (1977). Control of the wax moth *Galleria mellonella* on bee comb by H-sterotype V *Bacillus thuringiensis* and the effect of chemical additives. *Apidologie (Celle), 8*(2), 155-168.
[http://dx.doi.org/10.1051/apido:19770206]

Burges, H.D. (1978). Control of the wax moths: physical, chemical and biological methods. *Bee World, 59*(4), 129-138.
[http://dx.doi.org/10.1080/0005772X.1978.11097713]

Cantwell, G.E., Lehnert, T. (1968). Mortality of Nosema apis and the greater the wax moth, *Galleria mellonella* L., caused by heat treatment. *Am. Bee J., 108*, 56-57.

Cantwell, G.E., Smith, L.J. (1970). Control of the more significant the wax moth *Galleria mellonella*, in honeycomb and comb honey. *Am. Bee J., 10*, 141.

Cantwell, G.E., Matthenius, J.C. (1973). From the moon to the beehive. *Agric. Res. (US Dep. Agric.), 22*, 10.

Charles, A. (2017). FombongThe Biology and Control of the Greater The wax moth, *Galleria mellonella*. *Insects, 8*, 1-17.

Charriere, J.D., Imdorf, A. (1999). Protection of honeycombs from the wax moth damage. *Am. Bee J., 139*, 627-630.

Chase, R.W. (1921). The length of life of the larva of the wax moth, *Galleria mellonella* L., in its different stadia. *Trans. Wis. Acad. Sci. Arts Lett., 20*, 263-267.

Crane, E. (2000). Prevention and treatment of diseases and pests of honey bees: The world picture. *New*

Zealand Beekeeper., 10, 5-8.

Dayan, F.E., Cantrell, C.L., Duke, S.O. (2009). Natural products in crop protection. *Bioorg. Med. Chem.,* 17(12), 4022-4034.
[http://dx.doi.org/10.1016/j.bmc.2009.01.046] [PMID: 19216080]

Desalegne, B. *Origin and characterization of honeybee (A. melifera) pollen source around Utrecht University, The Netherlands (Doctoral dissertation*Utrecht University, Faculty of Biology, Department of behavioural social inset unit.

Dietemann, V., Nazzi, F., Martin, S.J., Anderson, D.L., Locke, B., Delaplane, K.S., Wauquiez, Q., Tannahill, C., Frey, E., Ziegelmann, B., Rosenkranz, P. (2013). Standard methods for varroa research. *J. Apic. Res.,* 52(1), 1-54.

Dougherty, E.M., Cantwell, G.E., Kuchinski, M. (1982). Biological control of the greater the wax moth (Lepidoptera: Pyralidae), utilizing *in vivo* and *in vitro* propagated baculovirus. *J. Econ. Entomol.,* 75(4), 675-679.
[http://dx.doi.org/10.1093/jee/75.4.675]

Dubovskiy, I.M., Grizanova, E.V., Whitten, M.M.A., Mukherjee, K., Greig, C., Alikina, T., Kabilov, M., Vilcinskas, A., Glupov, V.V., Butt, T.M. (2016). Immuno-physiological adaptations confer wax moth *Galleria mellonella* resistance to *Bacillus thuringiensis. Virulence,* 7(8), 860-870.
[http://dx.doi.org/10.1080/21505594.2016.1164367] [PMID: 27029421]

Eischen, F.A., Dietz, A. (1987). Growth and survival of *Galleria mellonella* (Lepidoptera: Pyralidae) larvae fed diets containing honey bee-collected plant resins. *Ann. Entomol. Soc. Am.,* 80(1), 74-77.
[http://dx.doi.org/10.1093/aesa/80.1.74]

Ellis, A.M., Hayes, G.W. (2009). Assessing the efficacy of a product containing *Bacillus thuringiensis* applied to honey bee (Hymenoptera: Apidae) foundation as a control for *Galleria mellonella* (Lepidoptera: Pyralidae). *J. Entomol. Sci.,* 44(2), 158-163.
[http://dx.doi.org/10.18474/0749-8004-44.2.158]

Ellis, J.D., Graham, J.R., Mortensen, A. (2013). Standard methods for wax moth research. *J. Apic. Res.,* 52(1), 1-17.
[http://dx.doi.org/10.3896/IBRA.1.52.1.10]

El-Sinary, N.H., Rizk, S.A. (2007). Entomopathogenic fungus, Beauveria bassiana (Bals.) and gamma irradiation efficiency against the greater wax moth, Galleria melonella (L.). *Am-Eur J Sci Rese.,* 2(1), 13-18.

Euverte, G., Martouret, D. (1963). Contribution a l'emp!oi de *Bacillus thuringiensis* Berliner pour Ia preservation de Ia eire d'abeille contre Galleria mellonelia. *Apidologie (Celle),* 6(4), 267-276.
[http://dx.doi.org/10.1051/apido:19630403]

Ferreira, T.P., Oliveira, E.E., Tschoeke, P.H., Pinheiro, R.G., Maia, A.M.S., Aguiar, R.W.S. (2017). Potential use of Negramina (Siparuna guianensis Aubl.) essential oil to control wax moths and its selectivity in relation to honey bees. *Ind. Crops Prod.,* 109, 151-157.
[http://dx.doi.org/10.1016/j.indcrop.2017.08.023]

Gillard, G My Friend, The wax moth. *Amer. Bee J. Vol.,* 149(6), 559-562.

Gulati, R., Kaushik, H.D. (2004). Enemies of honeybees and their management—A review. *Agric. Rev. (Karnal),* 25(3), 189-200.

Heimpel, AM (1971). Safety of insect pathogens for man and vertebrates. *Burges, HD Microbial control of insects and mites.*

Heitor, F. (1960). Sensibilite vis-a-vis de *Bacillus thuringiensis* des insectes nuisibles aux ruches: *Galleria mellonella* L. et *Achroia grisella* Fabr. Trans,XI int. *Congr. Ent,* 845-849.

Jafari, R., Goldasteh, S., Afrogheh, S. (2010). Control of the wax moth *Galleria mellonella* L.(Lepidoptera: Pyralidae) by the male sterile technique (MST). *Arch. Biol. Sci.,* 62(2), 309-313.

James, R.R. (2011). Potential of ozone as a fumigant to control pests in honey bee (Hymenoptera: Apidae)

hives. *J. Econ. Entomol., 104*(2), 353-359.
[PMID: 21510179]

Johansen, C. (1962). Impregnated foundation for wax moth control. *Glean. Bee Cult., 90*(11), 682-684.

Johansen, C (1968). Control of the wax moth with comb foundation impregnated with *Bacillus thuringiensis. Bull. apic., 11*, 8.

Krebs, H.M. (1957). Ethylene-dibromide-death to the wax moth control. *Clearing in Bee Culture., 90*, 682-684.

Mabrouk, M.S., Mahbob, M.A. (2015). Effect of Different Coloured Light Traps on Captures and Controlling The wax moth (Lepidoptera: Pyralidae). *Egypt. Acad. J. Biol. Sci. A Entomol., 8*(2), 17-24.

McGaughey, W., Johnson, D. (1994). Influence of crystal protein composition of *Bacillus thuringiensis* strains on cross-resistance in Indian meal moths (Lepidoptera: Pyralidae). *J. Econ. Entomol., 87*, 535-540.

Millar, E.S. (1965). *Bacillus thuringiensis* of no value in controlling the lesser the wax moth in stored honeycombs. *N. Z. J. Agric., 110*, 63.

Neumann, P., Evans, J.D., Pettis, J.S., Pirk, C.W.W., Schäfer, M.O., Tanner, G., Ellis, J.D. (2013). Standard methods for small hive beetle research. *J. Apic. Res., 52*(4), 1-32.
[http://dx.doi.org/10.3896/IBRA.1.52.4.19]

Nielsen, R.A., Brister, D. (1977). The more significant the wax moth: Adult behaviour. *Ann. Entomol. Soc. Am., 70*(1), 101-103.
[http://dx.doi.org/10.1093/aesa/70.1.101]

Noosidum, A., Hodson, A.K., Lewis, E.E., Chandrapatya, A. (2010). Characterization of new entomopathogenic nematodes from Thailand: foraging behaviour and virulence to the greater the wax moth, *Galleria mellonella* L.(Lepidoptera: Pyralidae). *J. Nematol., 42*(4), 281-291.
[PMID: 22736860]

Paddock, F.B. (1918). *The beemoth or waxworm..* Texas Agricultural Experiment Station.
[http://dx.doi.org/10.5962/bhl.title.57185]

Popolizio, E.R., Pailhe, L.A. (1973). Storing combs in "wax-moth-safe" storage rooms. *Proceedings of the 24th International Apicultural Congress,* 382-383.

Ritter, W, Akratanakul, P Honey bee diseases and pests: a practical guide.

Sangramsinh, P.B., Bhat, N.S., Reddy, G.N., Magar, P.N. (2014). Performance of sex pheromone blends in trapping greater the wax moth *Galleria mellonella. J. Insect Sci. (Ludhiana), 27*(1), 139-143.

Shelton, A.M., Robertson, J.L., Tang, J.D., Perez, C., Eigenbrode, S.D., Preisler, H.K., Wilsey, W.T., Cooley, R.J. (1993). Resistance of diamondback moth (Lepidoptera: Plutellidae) to *Bacillus thuringiensis* subspecies in the field. *J. Econ. Entomol., 86*(3), 697-705.
[http://dx.doi.org/10.1093/jee/86.3.697]

Shimanuki, H., Knox, D.A. (1997). Summary of Control Methods.*Honey Bee Pests, Predators, & Disease.* Published by AI Root Company.

Smith, T.L. (1965). External morphology of the larva, pupa, and adult of the wax moth, *Galleria mellonella* L. *J. Kans. Entomol. Soc.,* 287-310.

Surendra, N.S., Bhushanam, M., Reddy, M.S. (2010). Efficacy of natural plant products, Azadirachta indica, Ocimum sanctum and Pongamia pinnata in the management of greater wax moth, *Galleria mellonella* L. under laboratory conditions. *J. Appl. Nat. Sci., 2*(1), 5-7.
[http://dx.doi.org/10.31018/jans.v2i1.84]

Tabashnik, B.E., Finson, N., Johnson, M.W., Heckel, D.G. (1994). Cross-resistance to *Bacillus thuringiensis* toxin CryIF in the diamondback moth (Plutella xylostella). *Appl. Environ. Microbiol., 60*(12), 4627-4629.
[http://dx.doi.org/10.1128/aem.60.12.4627-4629.1994] [PMID: 16349471]

Tsegaye, A., Wubie, A.J., Eshetu, A.B., Lemma, M. (2014). Evaluation of different non-chemical wax moth

prevention methods in the backyards of rural beekeepers in the North West dry land areas of Ethiopia. *IOSR J. Agric. Vet. Sci., 7*(3), 29-36.
[http://dx.doi.org/10.9790/2380-07312936]

Vijayakumar, K.T., Neethu, T., Shabarishkumar, S., Nayimabanu Taredahalli, M.K., Bhat, N.S., Kuberappa, G.C. (2019). The survey, biology and management of greater the wax moth, *Galleria mellonella* L. in Southern Karnataka, India. *J. Entomol. Zool. Stud., 7*(4), 585-592.

Warren, L.O., Huddleston, P. (1962). Life history of the greater the wax moth, *Galleria mellonella* L., in Arkansas. *J. Kans. Entomol. Soc., 35*(1), 212-216.

Williams, J.L. (1997). Insects: Lepidoptera (moths).*Honey bee pests, predators, and diseases.* (pp. 121-141). Ohio, USA: The AI Root Company.

CHAPTER 7

Plastic Biodegradation by the Wax Moth: A Viable Alternative

Abstract: Environmental pollution due to plastic is becoming a concentration, drawing concern throughout the world. The wax moth larvae possess the potential for biodegradation of different types of plastic with or without the involvement of the intestinal microbiome in the larval gut. Similarly, mealworms and *Tenebrio molitor* have been reported to cause the degradation of polyethene and polystyrene mixtures. According to scientific literature, superworms such as Zophobas atratus can cause polystyrene degradation. The plastic is biodegradable with many bacterial genera, including *Pseudomonas, Ralstonia, Stenotrophomonas, Rhodococcus, Staphylococcus, Streptomyces, Bacillus*, *Aspergillus, Cladosporium, Penicillium*, and others. A few other invertebrates with complex gut microbiomes also possess this property of plastic biodegradation.

Keywords: Plastic degradation, The wax moth.

INTRODUCTION

The wax moth belongs to the Insecta Class, the Order Lepidoptera, the Superfamily Pyraloidea, and the Family Pyralidae (snout moths). *Achroia grisella* and *Galleria mellonella* are commercially bred. The wax worms have medium-sized white bodies with small black or brown heads. Under natural conditions, the wax moth lives in the bee colony as a nest parasite, feeding on the bee wax, cocoons, pollen, and exivae. Although the wax moth is considered a pest, some of its benefits include: the wax moth serves as an essential food source for many insectivorous animals and plants. The wax moth also serves as food for humans, some pet birds, and terrarium pets, as it possesses high-fat content, ability to rearingand can survive at low temperatures for weeks.

Additionally, the wax moth serves as food for bearded dragons (species in the genus Pogona), the neon tree dragon (*Japalura splendida*), geckos, brown anole (*Anolis sagrei*), turtles such as the three-toed box turtle (*Terrapene carolina triunguis*), and chameleons. The wax moth also serves as food for *Ceratophrys* frogs, Strauch's spotted newt (*Neurergus strauchii*), salamanders, domesticated hedgehogs, the greater honeyguide, assassin bugs, the genus Platymeris, and bluegills (*Lepomis macrochirus*).

The wax worm can be used as bait in case of fishing. Bait is used to attract and catch fish by attaching it to a fishing hook. The wax moth is used to capture Masu salmon (*Oncorhynchus masou*), white-spotted char (*Salvelinus leucomaenis*), and rainbow trout (*Oncorhynchus mykiss*). The caterpillars of the wax moth are known as wax worms.

The wax worms are used as an experimental model for animal research to examine the virulence mechanism of bacterial and fungal pathogens (Antunes *et al.,* 2011) because of the similarity in the innate immune system of mammals (Kavanagh *et al.,* 2004). The wax moth is an excellent experimental model because of its small size, accessible laboratory rearing protocol and handling, economical rearing cost, high biotic potential, and adaptability to laboratory conditions. In addition, the wax moth facilitated the screening of various bacteria and fungal strains, identifying genetic elements involved in pathogenesis. The wax moth larvae also possess the potential for biodegradation of polyethene (Yang *et al.,* 2014); Jambeck *et al.,* 2015; Bombelli *et al.,* 2017; Ball, 2017; Geyer *et al.,* 2017; Yang *et al.,* 2019; Cassone *et al.* 2020).

PLASTIC DEGREDATION

Different Types of Plastics

Plastic is a non-biodegradable synthetic polymer which can be derived from fossil oil. Plastics have been introduced since the 1960s. Plastics comprise carbon, hydrogen, oxygen, and chloride (Wu *et al.,* 2017). Further, plastics like polyethene [PE], polypropylene, and polystyrene, petroleum-derived synthetic polymers, possess hydrocarbon chains of different sizes and structures. Plastic is becoming a significant cause of environmental pollution, and about 92% of plastic is of the polyethene (PE) and polypropylene (PP) types that are exponentially used in packing. Different types of plastics contain additional solubilisers and other agents for the enhancement of their mechanical and physical properties. Excessive use of plastic has increased the challenge of plastic disposal and has resulted in environmental pollution. Therefore, it has become urgent to find ways for the biodegradation of plastic. The non-biodegradable nature of these chemicals results in the bioaccumulation of these chemicals (Barnes *et al.,* 2009; Rustagi *et al.,* 2011). "Weathering" and "photodegradation" are considered major methods for the initial biodegradation of plastics and result in modification of the plastics' chemical, physical, and mechanical properties.

PE is the most common petroleum-based plastic that is in excessive use in everyday life. Because of the high durability and high usage time of PE, it results in rapid accumulation, eventually causing environmental pollution and deterioration (Ammala *et al.,* 2011; Roy *et al.,* 2011; Shah *et al.,* 2008; Zettler *et al.,* 2013). PE is decomposed into low molecular weight substances like alkanes, alkenes, ketones, aldehydes, various alcohols, and fatty acids (Albertsson *et al.,* 1987, 1998; Tokiwa *et al.,* 2009). The decomposition of plastic in multiple environments has been investigated to find out the solution to plastic degradation (Jones *et al.,* 1974; Albertsson and Karlsson, 1988; Pegram and Andrady, 1989; Ohtake *et al.,* 1998; Nowack and Bucheli, 2007; Artham *et al.,* 2009; Nkwachukwu *et al.,* 2013). The bee wax comprises palmitoleate, long-chain aliphatic alcohols, and hydrocarbons.

Different Test Models for the Plastic Degradation

Considering the challenge of accumulating plastic, there is an increasing demand for materials with biodegradable characteristics. Various research explorations are procurable in the scientific literature that witnesses the list of some insects of Coleoptera and Lepidoptera that voluntarily consume plastic and produce byproducts during the biodegradation process. Different insects include the greater wax moth (GWM), the lesser wax moth (LWM), *Plodia interpunctella,etc.*

The Wax Moth and Plastic

Galleria mellonella, Achoria grisella, and *Plodia interpunctella* have been identified for their ability to eat and digest polyethene bags (Figs. **7a-d**). The wax worm metabolises polyethene to ethylene glycol, which possesses the capacity for rapid biodegradation (Charles *et al.,* 2017). Bombelli *et al.* (2017) discovered that 100 *Galleria mellonella* wax worms could biodegrade nearly 0.1 gramme (0.0032 ounces) of plastic in 12 hours. The wax moth's ability to degrade plastic has been reported for the first time by Federica Bertocchini of the Institute of Biomedicine and Biotechnology of Cantabria in Spain while working on honey bees. She noted that the wax worms had consumed the plastic bag within which they were contained. Further, she reported with her team that about 100 wax worms consumed a polyethene shopping bag within 40 minutes. To check whether the wax worm just eats up the polyethene bag or digests it, they just smashed the wax worm and spread it on the plastic sheet. About 13% of plastic disappears after 14 hours. That indicates that wax worms are capable of digesting plastic. Further, they reported the presence of ethylene glycol as the end product of plastic digestion, which confirms the ability of the wax moth for plastic biodegradation (Khyade *et al.* 2018).

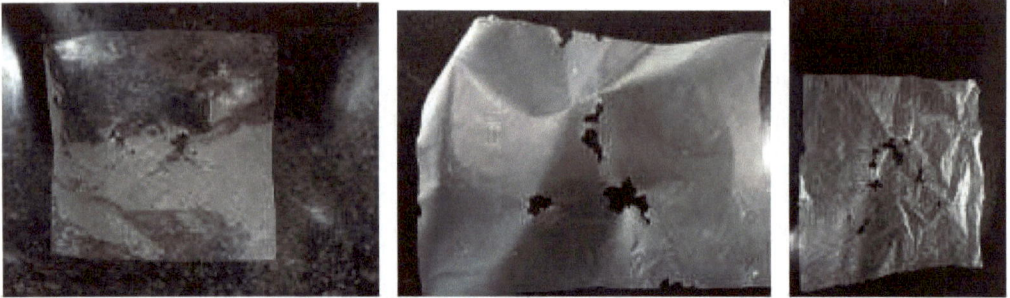

Fig. (7a, b, c). The greater wax moth biodegrades low-density polyethene samples in the laboratory.

Fig. (7d). The wax moth larvae feed on low-density polyethene in a beaker.

Kong *et al.,* 2019 reported that if *Galleria mellonella* is fed with beeswax or wheat bran as a co-diet, that will enhance its capability to degrade plastic. Furthermore, they reported that the GWM impairs approximately 0.88 and 1.95 g of PS and PE plastic over 21 days. Further, they said that if the bee wax was provided to wax worms along with plastic, the consumption of plastic decreased thereafter. The presence of bee wax and bran influences the core gut microbiome of the larvae feeding on PE and PS, which indicates that supplementing co-diet affects the physiological properties of larvae, hence the plastic biodegradation capacity.

Wax Moth (WM) Enzymes Involved

Kong *et al.,* 2019 assembled genomic and transcriptomic data from *G. mellonella* to find the long-chain hydrocarbon biodegradation mechanism. Certain enzymes like carboxylesterase, lipase and fatty-acid-metabolism-related enzymes are transcribed excessively when the GWM is fed on bee wax.

Christope *et al.,* 2020 tried to correlate RNA sequencing with biochemical approaches to evaluate the role of the wax moth in plastic degradation. It has been reported that the wax moth larvae fed on plastic exhibit high fatty acid metabolism. Further, they concluded that after feeding wax worms on plastic, there had been an increase in the activity of alcohol dehydrogenase. In addition, they observed that transcriptome analysis of the GWM indicated carboxylesterase and lipase 1 and 3 family-related genes were overexpressed in the fat body cells and the gut cells of groups fed on nutrition-rich foo. Other related enzymes which further help in degradation include acetyl-coenzyme A (acyl-CoA) dehydrogenase, enoyl-CoA hydrolase, 3-hydroxy acyl-CoA dehydrogenase, and 3-ketoacyl-CoA thiolase.

Involved Micro-organism of Wax Moth

Several bacterial strains have been identified in the wax moth, including *Lysinibacillus fusiformis, Bacillus aryabhattai, Acinetobacter,* and *Micro-bacterium oxydans* (Rajandas *et al.* 2012; Bombelli *et al.* 2017; Ren *et al.* 2019; Cassone *et al.* 2020; Montazer *et al.,* 2020).

Kong *et al.* (2019) used the GWM to investigate wax degradation with and without microbes. They have demonstrated that the biodegradation of plastic occurs even in the absence of gut micro-organisms. Further, they constructed a genomic reference map of the GWM, which helps identify responsible genetic elements for the degradation of long-chain hydrocarbons. Transcriptomics of intestinal micro-organisms helps identify bacterial enzymes responsible for short-chain degradation. The Genomics and Proteomics of the GWM assist in developing the biological degradation of PE methods. Even in the absence of gut microbes, biodegradation end products can be detected, which confirms the role of the wax moth in plastic degradation (Kong *et al.,* 2019).

Peydaei *et al.,* 2020 reported the wax worm's role in the salivary gland in PE degradation. According to them, the surface of PE gets affected due to the chewing of wax worms, and they have also detected degradation intermediates like carbonyl groups. They analysed that salivary gland proteins get changed on the consumption of PE. Further, they found that after the consumption of PE, wax worms carry out biodegradation of PE by activating associated enzymes for fatty acid beta-oxidation. The GWM can carry out biodegradation of PE even independent of the intestinal microbiota, which could play a secondary role in the degradation process (Kong *et al.,* 2019). Jacquin *et al.,* 2019 hypothesised that microbiota could carry out synergetic roles in PE degradation.

Peydaei *et al.,* 2020 reported that when the GWM is fed with the LDPE, observable holes appear on the sheet after approximately 40 min. They said that wax worms had created large holes (6.3 mm) and pits (14 μm). FTIR of normal PE indicates the presence of 2919, 2851, 1473, 1463, and 1377 and around 730 cm-1 peaks (Gulmine *et al.,* 2002). But after feeding PE by WM, they confirmed the presence of 3300 cm−1 corresponding to the OH-stretching peak of ethylene glycol in the frass of the insect. They reported about 1439 proteins across all samples in the salivary glands. Further, they identified about 47 enzymes in the salivary gland tissue related to metabolism, which includes oxidoreductases such as (3R)-3-hydroxy acyl-CoA dehydrogenase, aldehyde dehydrogenase, and catalase; transferases including acetyl-CoA C-acyltransferase and glutathione transferase; and hydrolases such as multiple inositol-polyphosphate phosphatases.

The Mechanism of Plastic Degradation by the Wax Moth

Bee wax comprises hydroxypalmitate monoesters with long-chain alcohols, diesters, triesters, hydroxy polyesters, acidic polyesters, free fatty acids, free alcohols, and some unidentified compounds (Maia and Nunes, 2012). PE consists of a backbone composed of long carbon chains similar to beeswax (Kundungal *et al.,* 2019a). Kong *et al.* (2019) hypothesised that long-chain hydrocarbons are depolymerised or hydrolysed by the host (WM), and thereafter long-chain fatty acids are metabolised by gut microbiota. Furthermore, they reported that carboxylesterases, lipases, and other enzymes are involved in the oxidation of fatty acids in the GWM.

In the case of the wax moth, in long-chain hydrocarbons, hydroxyl groups are introduced *via* monooxygenase reactions catalysed by CYP enzymes (Ortiz, 2010). The CYP9 gene family is expanded in the wax moth (Ortiz, 2010). After monooxygenase-catalysed hydrolysis by CYPs, the resulting alcohol is subjected to alcohol oxidation and beta-oxidation, which results in the production of butanoyl-CoA and propionyl-CoA byproducts. Finally, these end products result in acetyl-CoA generation by butanoate and propanoate metabolism (Gatter *et al.,* 2014). Houten and Wanders (2010) reported that in wax worms, expansion of the fatty acid metabolism, fatty acid biosynthesis, and fatty acid degradation pathways indicate the metabolism of wax into $C_{15}H_{31}COONa$ short-chain fatty acids (Houten and Wanders, 2010; Kong *et al.,* 2019). PE film has been covered with smashed wax worms to exclude the mechanical action of the masticatory system. Further, gravimetric analysis confirmed a significant loss of 13% of PE over 14 hours. FTIR of untreated samples indicated only two peaks of 2,921 and 2,852 cm-1 which are considered the classical signature of PE, whereas after treatment with worm homogenate, an additional peak of 3,350 cm-1 was seen.

Furthermore, they confirmed that intestinal microbiota degrades short-chain fatty acid wax products but not long-chain hydrocarbon wax. The investigation into PE biodegradation by GWM indicated that biodegradation of PE occurs with partial interaction with gut micro-organisms. They analysed the presence of octadecanoic acid, cis-vaccenic acid, and palmitic acid in the frass of beeswax-fed G. mellonella lacking intestinal microbiota.

Ren *et al.,* 2019 analysed PE plastic degradation by micro-organisms from the gut of the wax moth (*Galleria mellonella*) with the help of scanning electron microscopy (SEM) coupled with energy dispersive spectroscopy (EDS), atomic force microscopy (AFM), Fourier transforms infrared spectroscopy (FTIR), and liquid chromatography-tandem mass spectrometry (LC-MS).

Kundungal *et al.* (2019) studied the preactivation of PE for biodegradation with solar radiation. Furthermore, they analysed pretreated LDPE (PTLDPE) and untreated LDPE (UTLDPE) with AFM, FTIR, and 1H NMR techniques. PE film has been covered with smashed wax worms to exclude the mechanical action of the masticatory system. Furthermore, gravimetric analysis confirmed a significant loss of 13% of PE over 14 hours. Furthermore, FTIR of untreated samples indicated only two peaks of 2,921 and 2,852 cm-1 which are considered the classical signature of PE, whereas, after treatment with worm homogenate, an additional peak of 3,350 cm-1 was seen.

Solar pretreatment enhances surface roughness, which favours the wax worm intake of PE. Many reports indicate that PE can be degraded with the help of bacteria, fungi, and microbes. Molecular weight, physical and chemical nature, presence and/or absence of -OH and -CO in the polymer, and the ability of micro-organisms to produce extracellular enzymes into low molecular weight fragments all influence microbial degradation efficiency (Novotn *et al.,* 2018). Exploration indicates that enzymes like laccase and alkane hydroxylase help in extracellular digestion (Santo *et al.,* 2013; Jeon and Kim, 2015). Further activation can be given by UV radiation, thermal oxidation, pro-oxidant additives, photocatalysis, *etc.*, which induce synergetic effects on the biodegradation of plastic to small fragments (Ro *et al.,* 2008; Liu *et al.,* 2009; Thomas *et al.,* 2013; Mehmood *et al.,* 2016). The biodegradation of PE with the help of the wax moth *Galleria mellonella* provides a new insight into this area of research (Bombelli *et al.,* 2017). Bee wax is composed of a highly diverse mixture of lipids, including alkenes, alkanes, esters, and fatty acids (Bombelli *et al.*, 2017).

A diet composed of LDPE alone is not sufficient for the growth of larvae (Billen *et al.*, 2020). PE biodegradation by micro-organisms has some drawbacks, including the production of culture medium, high maintenance costs, and the generation of microplastics (Billen *et al.* 2020).

MICRO-ORGANISMS FOR PLASTIC DEGRADATION

Biodegradation of PE occurs mainly due to the activity of various micro-organisms after thermal oxidation (Albertsson *et al.*, 1987; Tokiwa *et al.*, 2009).

Polyethene degradation occurs with the treatment of nitric acid and with incubation with the fungus *Penicillium simplicissimum* (Yamada-Onodera *et al.*, 2001). In contrast, *Nocardia asteroids* induce slow PE degradation within 4–7 months of exposure (Bonhomme *et al.*, 2003). In both the cases cited above, after Fourier transform infrared spectroscopy (FTIR) analysis, treated samples revealed the formation of absorbance of around 3,300 cm-1. Poly (ethylene terephthalate) (PET) degradation with the help of an isolated bacterium, Ideonella sakaiensis, has been reported by Yoshida *et al.*, 2016.

Polyethene is a synthetic ethylene polymer available in two forms: high-density (HD-PE) or low-density (LD-PE) polyethene. Polyethene is used as packaging material, and more than 90 genera of bacteria and fungi possess the ability to decompose plastics (Mahdiyah and Mukti, 2013). The degradation of long-chain hydrocarbons is an important step in the biodegradation of PE.

Micro-organisms responsible for plastic degradation are present in all habitats, including terrestrial, freshwater, and marine environments (Urbanek *et al.*, 2018; Morohoshi *et al.*, 2018; Raddadi and Fava, 2019). Further, the biodegradation of plastics is influenced by properties of the plastic polymers like molecular weight and density. Gilliam *et al.*, 1989 reported micro-organisms such as P. purpurogenum, A. flavus group, A. niger, and others in wax moth frass.

In *Plodia interpunctella* wax worms' gut, two bacterial strains, *Enterobacter asburiae* and *Bacillus sp* are responsible for the degradation of polyethene. These two bacterial strains can make plastic film hydrophobic and create pits and cavities of 0.3-0.4 μm in depth, which can be observed under a microscope.

POSSIBLE MECHANISM

These caterpillars do biodegradation of LDPE and excrete ethylene glycol as a byproduct, as detected through FTIR. Bee wax is composed of long fatty esters and hydrocarbons. Therefore, it is likely for the wax moth to have a unique microbiota for the digestion of complex foods. The general pathway for LDPE degradation involves several oxidases and hydrolases for the reduction of molecular weight and oxidation. The specific degeneration results in the yielding of carbon-rich substrate for β--oxidation, the citric acid cycle, and other byproducts (Wilkes and Aristilde, 2017). It has been reported that in the greater wax moth, these microbiomes are solely responsible for biodegradation (Cassone *et al.*, 2020; Billen *et al.*, 2020). It has been observed that GWM can do biodegradation of PE even without a gut microbiome (Kong, 2019). Furthermore, they concluded that genetic elements required for the initial breakdown and for oxidation are abundant in the wax moth. They reported that PE-fed caterpillars, concomitantly with the normal function of the digestive process, maintain high fatty acid metabolism.

Wax worms fed on LDPE maintain high lipid reserves, which larvae can utilise for development. Further, these larvae have been assessed to be equipped with the enzymes alcohol dehydrogenase and lactate dehydrogenase, which enhance their ability to degrade long-chain fatty alcohols. Both of these enzymes help to convert long-chain fatty alcohol into aldehyde and carboxylate. They experimented with 100 larvae of GWM, which fed on LDPE for 72%. They discovered that 84% of the larvae were provided within the first 24 hours, 64% consumed plastic at 48 hours, and 37% finished plastic at 72 hours. Further, they concluded that each feeding waxworm consumed during the first day was about 250 µg on the very first day, which had been reduced thereafter. They discovered that in PE-fed larvae, transcripts for fatty acid metabolism, such as hydrolases, desaturases, lipases, and those involved in -oxidation, are up-regulated (Harwood, 1988; Christophe *et al.*, 2020).

For microbial degradation, these synthetic compounds are degraded with exoenzymes to multimers and dimers, which can be incorporated into microbial cells. After that, these products enter into classical degradation pathways to yield energy.

ANALYSIS OF INFORMATION

Further, analysis with liquid chromatography coupled with mass spectrometry (HPLC–MS) indicated that in PE waxworm smash treated sheets, and three new peaks appeared at the lower end of the m/z region (110.0, 122.9, and 170.0). Further, through Atomic Force Microscopy (AFM), a change in the topography of the PE surface has been reported, which indicates that in waxworm, smashed sheet surface roughness increases by 140%. Beewax is made up of alkanes, alkenes, fatty acids, and esters, with the most common hydrocarbon bond being

CH2-CH2, as in PE. FTIR detection of the breakdown of the C-C single bond in bee wax and treated PE is standard (Bombelli *et al.*, 2017). Christophe *et al.*, 2020 reported that the gut transcript of PE-fed larvae indicated a molecular signature of enhanced fatty acid metabolism in addition to normal intestinal function. Further, they reported that the activity of enzymes involved in lipid oxidation is relatively high in PE-fed larvae and concluded that there is a similarity in the chemical structure of PE and bee wax, which endows waxworm with an extra-ordinary ability to biodegrade polyethene, polypropylene, and polystyrene. A few insects of the orders Coleoptera and Lepidoptera can biodegrade plastic. *Tenebrio molitor* and *Tenebrio obscurus* (yellow and dark mealworms), *Achroia grisella* (lesser waxworm), and *Galleria mellonella* (greater wax moth) can biodegrade polystyrene, polyethene, and low-density polyethene (LDPE) (Yang *et al.*, 2015; Brandon *et al.*, 2018; Bombelli *et al.*, 2017; Peng *et al.*, 2019; Kundungal *et al.*, 2019).

The wax moth gut contains *Enterobacter* sp. D1, which helps in the simplification of PE. These bacterial treatments induce roughness, depression, and cracks on the surface of plastic film as detected by scanning electron microscopy (SEM) and atomic force microscopy (AFM). Further, FTIR indicates the presence of carbonyl functional and ether groups on the PE film. Liquid chromatography-tandem mass spectrometry (LC-MS) indicated alcohols, esters, and acids when exposed to *Enterobacter* sp. Microbial degradation can occur in four steps, including microbial cells after growth on plastic sheets produce hydrophilic groups and long-chain hydrocarbons get oxidative to produce short chains, which are further broken down into fatty acids. Fatty acid gets oxidised into H_2O, CO_2, and humus (Shah *et al.*, 2008; Singh and Sharma, 2008). *Bacillus, Pseudomonas, Staphylococcus, Streptococcus, Streptomyces, Brevibacterium, Nocardia, Moraxella, Penicillium,* and *Aspergillus* can all cause plastic biodegradation (Jones *et al.*, 1974; Pegram *et al.*, 1989; Restrepo-Flórez *et al.*, 2014; Krueger *et al.*, 2015). Certain fungi, like *Penicillium simplicissimum* and *Nocardia* asteroids, also cause plastic degradation (Yamada-Onodera, 2000). According to Yang Jun *et al.*, plastic degradation occurs with the help of two strains, including *Enterobacter asburiae* YT1 and *Bacillus sp.* YP1. These fungi contribute up to 6% and 11% of the degradation of PE films, respectively (Yang *et al.*, 2014).

CONCLUSION

The caterpillar of the wax moth is a plastivore, which possesses the capability to degrade polyethene at unprecedented rates. Free and gut-inhibiting micro-organisms facilitate the degradation of low-density polyethene (LDPE) and glycol production as an end product. The caterpillar fed on a natural honeycomb diet exhibits a greater abundance of microbes than a starved caterpillar.

REFERENCES

Antunes, L.C.S., Imperi, F., Carattoli, A., Visca, P. (2011). Deciphering the multifactorial nature of Acinetobacter baumannii pathogenicity. *PLoS One,* *6*(8), e22674.
[http://dx.doi.org/10.1371/journal.pone.0022674] [PMID: 21829642]

Austin, H.P., Allen, M.D., Donohoe, B.S., Rorrer, N.A., Kearns, F.L., Silveira, R.L., Pollard, B.C., Dominick, G., Duman, R., El Omari, K., Mykhaylyk, V., Wagner, A., Michener, W.E., Amore, A., Skaf, M.S., Crowley, M.F., Thorne, A.W., Johnson, C.W., Woodcock, H.L., McGeehan, J.E., Beckham, G.T. (2018). Characterization and engineering of a plastic-degrading aromatic polyesterase. *Proc. Natl. Acad. Sci. USA,* *115*(19), E4350-E4357.
[http://dx.doi.org/10.1073/pnas.1718804115] [PMID: 29666242]

Ball, P. (2017). Plastics on the menu. *Nat. Mater.,* *16*(6), 606.
[http://dx.doi.org/10.1038/nmat4912] [PMID: 28541312]

Barnes, D.K.A., Galgani, F., Thompson, R.C., Barlaz, M. (2009). Accumulation and fragmentation of plastic debris in global environments. *Philos. Trans. R. Soc. Lond. B Biol. Sci.,* *364*(1526), 1985-1998.
[http://dx.doi.org/10.1098/rstb.2008.0205] [PMID: 19528051]

Barth, A. (2007). Infrared spectroscopy of proteins. *Biochim. Biophys. Acta Bioenerg.,* *1767*(9), 1073-1101.
[http://dx.doi.org/10.1016/j.bbabio.2007.06.004] [PMID: 17692815]

Billen, P., Khalifa, L., Van Gerven, F., Tavernier, S., Spatari, S. (2020). Technological application potential of polyethylene and polystyrene biodegradation by macro-organisms such as mealworms and wax moth larvae. *Sci. Total Environ.,* *735*, 139521.
[http://dx.doi.org/10.1016/j.scitotenv.2020.139521] [PMID: 32470676]

Bombelli, P., Howe, C.J., Bertocchini, F. (2017). Polyethylene bio-degradation by caterpillars of the wax moth *Galleria mellonella*. *Curr. Biol.,* *27*(8), R292-R293.
[http://dx.doi.org/10.1016/j.cub.2017.02.060] [PMID: 28441558]

Bonhomme, S., Cuer, A., Delort, A-M., Lemaire, J., Sancelme, M., Scott, G. (2003). Environmental biodegradation of polyethylene. *Polym. Degrad. Stabil.,* *81*(3), 441-452.
[http://dx.doi.org/10.1016/S0141-3910(03)00129-0]

Brandon, A.M., Gao, S.H., Tian, R., Ning, D., Yang, S.S., Zhou, J., Wu, W.M., Criddle, C.S. (2018). Biodegradation of polyethene and plastic mixtures in mealworms (Larvae of *Tenebrio molitor*) and effects on the gut microbiome. *Environ. Sci. Technol.,* *52*(11), 6526-6533.
[http://dx.doi.org/10.1021/acs.est.8b02301] [PMID: 29763555]

Cassone, B.J., Grove, H.C., Elebute, O.O., Villanueva, S.M., LeMoine, C.M.R. (2020). Role of the intestinal microbiome in polyethene degradation by larvae of the greater the wax moth (*Galleria mellonella*). *Proc. Biol. Sci.,* *287*, 20200112.
[http://dx.doi.org/10.1098/rspb.2020.0112] [PMID: 32126962]

Christophe, MR, Harald, C.G, Charlotte, M.S, Bryan, J.C (2020). A Very Hungry Caterpillar: Polyethylene Metabolism and Lipid Homeostasis in Larvae of the Greater The wax moth (*Galleria mellonella*).
[http://dx.doi.org/10.1021/acs.est.0c04386]

Geyer, R., Jambeck, J.R., Law, K.L. (2017). Production, use, and fate of all plastics ever made. *Sci. Adv.,* *3*(7), e1700782.
[http://dx.doi.org/10.1126/sciadv.1700782] [PMID: 28776036]

Gilliam, M., Prest, D.B., Lorenz, B.J. (1989). Microbes from apiarian sources: Molds in frass from larvae of the greater wax moth, *Galleria mellonella*. *J. Invertebr. Pathol.,* *54*(3), 406-408.
[http://dx.doi.org/10.1016/0022-2011(89)90126-2]

Kundungal, Harsha, Gangarapu, Manjari, Sarangapani, Saran, Patchaiyappan, Arunkumar, Devipriya, Suja Purushothaman Role of Pretreatment and Evidence for the Enhanced Biodegradation and Mineralisation of Low-Density Polyethylene Films by Greater Waxworm.
[http://dx.doi.org/10.1080/09593330.2019.1643925]

Harwood, J.L. (1988). Fatty acid metabolism. *Annu. Rev. Plant Physiol. Plant Mol. Biol., 39*(1), 101-138.
[http://dx.doi.org/10.1146/annurev.pp.39.060188.000533]

Jacquin, J., Cheng, J., Odobel, C., Pandin, C., Conan, P., Pujo-Pay, M., Barbe, V., Meistertzheim, A.L., Ghiglione, J.F. (2019). Microbial ecotoxicology of marine plastic debris: a review on colonisation and biodegradation by the "plastisphere". *Front. Microbiol., 10*, 865.
[http://dx.doi.org/10.3389/fmicb.2019.00865] [PMID: 31073297]

Jambeck, J.R., Geyer, R., Wilcox, C., Siegler, T.R., Perryman, M., Andrady, A., Narayan, R., Law, K.L. (2015). Plastic waste inputs from land into the ocean. *Science, 347*(6223), 768-771.
[http://dx.doi.org/10.1126/science.1260352] [PMID: 25678662]

Jeon, H.J., Kim, M.N. (2015). Functional analysis of alkane hydroxylase system derived from *Pseudomonas* aeruginosa E7 for low molecular weight polyethylene biodegradation. *Int. Biodeterior. Biodegradation, 103*, 141-146.
[http://dx.doi.org/10.1016/j.ibiod.2015.04.024]

Jones, P.H., Prasad, D., Heskins, M., Morgan, M.H., Guillet, J.E. (1974). Biodegradability of photodegraded polymers. I. Development of experimental procedures. *Environ. Sci. Technol., 8*(10), 919-923.
[http://dx.doi.org/10.1021/es60095a010]

Khyade, V.B. (2018). Review on biodegradation of plastic through waxworm (Order: Lepidoptera; Family: Pyralidae). *International Academic Journal of Economics, 5*(1), 84-91.
[http://dx.doi.org/10.9756/IAJE/V5I1/1810008]

Kong, H.G., Kim, H.H., Chung, J., Jun, J., Lee, S., Kim, H.M., Jeon, S., Park, S.G., Bhak, J., Ryu, C.M. (2019). The *Galleria mellonella* hologenome supports the microbiota-independent metabolism of long-chain hydrocarbon beeswax. *Cell Rep., 26*(9), 2451-2464.e5.
[http://dx.doi.org/10.1016/j.celrep.2019.02.018] [PMID: 30811993]

Krueger, M.C., Harms, H., Schlosser, D. (2015). Prospects for microbiological solutions to environmental pollution with plastics. *Appl. Microbiol. Biotechnol., 99*(21), 8857-8874.
[http://dx.doi.org/10.1007/s00253-015-6879-4] [PMID: 26318446]

Kundungal, H., Gangarapu, M., Sarangapani, S., Patchaiyappan, A., Devipriya, S.P. (2019). Efficient biodegradation of polyethylene (HDPE) waste by the plastic-eating lesser waxworm (*Achroia grisella*). *Environ. Sci. Pollut. Res. Int., 26*(18), 18509-18519.
[http://dx.doi.org/10.1007/s11356-019-05038-9] [PMID: 31049864]

Kwadha, C.A., Ong'amo, G.O., Ndegwa, P.N., Raina, S.K., Fombong, A.T. (2017). The biology and control of the greater wax moth, *Galleria mellonella*. *Insects, 8*(2), 61.
[http://dx.doi.org/10.3390/insects8020061] [PMID: 28598383]

Liu, G.L., Zhu, D.W., Liao, S.J., Ren, L.Y., Cui, J.Z., Zhou, W.B. (2009). Solid-phase photocatalytic degradation of polyethylene–goethite composite film under UV-light irradiation. *J. Hazard. Mater., 172*(2-3), 1424-1429.
[http://dx.doi.org/10.1016/j.jhazmat.2009.08.008] [PMID: 19716230]

Maia, M., Nunes, F.M. (2013). Authentication of beeswax (*Apis mellifera*) by high-temperature gas chromatography and chemometric analysis. *Food Chem., 136*(2), 961-968.
[http://dx.doi.org/10.1016/j.foodchem.2012.09.003] [PMID: 23122150]

Mehmood, C.T., Qazi, I.A., Hashmi, I., Bhargava, S., Deepa, S. (2016). Biodegradation of low density polyethylene (LDPE) modified with dye sensitized titania and starch blend using *Stenotrophomonas* pavanii. *Int. Biodeterior. Biodegradation, 113*, 276-286.
[http://dx.doi.org/10.1016/j.ibiod.2016.01.025]

Montazer, Z., Habibi Najafi, M.B., Levin, D.B. (2021). *In vitro* degradation of low-density polyethylene by new bacteria from larvae of the greater wax moth, *Galleria mellonella. Can. J. Microbiol., 67*(3), 249-258.
[http://dx.doi.org/10.1139/cjm-2020-0208] [PMID: 33306436]

Montazer, Z., Habibi Najafi, M.B., Levin, D.B. (2020). Challenges with verifying microbial degradation of

polyethene. *Polymers (Basel), 12*(1), 123.
[http://dx.doi.org/10.3390/polym12010123] [PMID: 31948075]

Morohoshi, T., Oi, T., Aiso, H., Suzuki, T., Okura, T., Sato, S. (2018). Biofilm formation and degradation of commercially available biodegradable plastic films by bacterial consortiums in freshwater environments. *Microbes Environ., 33*(3), 332-335.
[http://dx.doi.org/10.1264/jsme2.ME18033] [PMID: 30158390]

Muhonja, C.N., Makonde, H., Magoma, G., Imbuga, M. (2018). Biodegradability of polyethylene by bacteria and fungi from Dandora dumpsite Nairobi-Kenya. *PLoS One, 13*(7), e0198446.
[http://dx.doi.org/10.1371/journal.pone.0198446] [PMID: 29979708]

Novotný, Č., Malachová, K., Adamus, G., Kwiecień, M., Lotti, N., Soccio, M., Verney, V., Fava, F. (2018). Deterioration of irradiation/high-temperature pretreated, linear low-density polyethylene (LLDPE) by *Bacillus amyloliquefaciens. Int. Biodeterior. Biodegradation, 132*, 259-267.
[http://dx.doi.org/10.1016/j.ibiod.2018.04.014]

Pegram, J.E., Andrady, A.L. (1989). Outdoor weathering of selected polymeric materials under marine exposure conditions. *Polym. Degrad. Stabil., 26*(4), 333-345.
[http://dx.doi.org/10.1016/0141-3910(89)90112-2]

Peng, B.Y., Su, Y., Chen, Z., Chen, J., Zhou, X., Benbow, M.E., Criddle, C.S., Wu, W.M., Zhang, Y. (2019). Biodegradation of polystyrene by dark (*Tenebrio obscurus*) and yellow (*Tenebrio molitor*) mealworms (Coleoptera: Tenebrionidae). *Environ. Sci. Technol., 53*(9), 5256-5265.
[http://dx.doi.org/10.1021/acs.est.8b06963] [PMID: 30990998]

Peydaei, A., Bagheri, H., Gurevich, L., de Jonge, N., Nielsen, J.L. (2020). Impact of polyethylene on salivary glands proteome in Galleria melonella. *Comp. Biochem. Physiol. Part D Genomics Proteomics, 34*, 100678.
[http://dx.doi.org/10.1016/j.cbd.2020.100678] [PMID: 32163748]

Raddadi, N., Fava, F. (2019). Biodegradation of oil-based plastics in the environment: Existing knowledge and needs of research and innovation. *Sci. Total Environ., 679*, 148-158.
[http://dx.doi.org/10.1016/j.scitotenv.2019.04.419] [PMID: 31082589]

Rajandas, H., Parimannan, S., Sathasivam, K., Ravichandran, M., Su Yin, L. (2012). A novel FTIR-ATR spectroscopy based technique for the estimation of low-density polyethylene biodegradation. *Polym. Test., 31*(8), 1094-1099.
[http://dx.doi.org/10.1016/j.polymertesting.2012.07.015]

Ren, L., Men, L., Zhang, Z., Guan, F., Tian, J., Wang, B., Wang, J., Zhang, Y., Zhang, W. (2019). Biodegradation of polyethene by Enterobacter sp. D1 from the guts of the wax moth *Galleria mellonella. Int. J. Environ. Res. Public Health, 16*(11), 1941.
[http://dx.doi.org/10.3390/ijerph16111941] [PMID: 31159351]

Restrepo-Flórez, J.M., Bassi, A., Thompson, M.R. (2014). Microbial degradation and deterioration of polyethylene – A review. *Int. Biodeterior. Biodegradation, 88*, 83-90.
[http://dx.doi.org/10.1016/j.ibiod.2013.12.014]

Roy, P.K., Titus, S., Surekha, P., Tulsi, E., Deshmukh, C., Rajagopal, C. (2008). Degradation of abiotically aged LDPE films containing pro-oxidant by bacterial consortium. *Polym. Degrad. Stabil., 93*(10), 1917-1922.
[http://dx.doi.org/10.1016/j.polymdegradstab.2008.07.016]

Rustagi, N., Singh, R., Pradhan, S.K. (2011). Public health impact of plastics: An overview. *Indian J. Occup. Environ. Med., 15*(3), 100-103.
[http://dx.doi.org/10.4103/0019-5278.93198] [PMID: 22412286]

Santo, M., Weitsman, R., Sivan, A. (2013). The role of the copper-binding enzyme – laccase – in the biodegradation of polyethylene by the actinomycete *Rhodococcus* ruber. *Int. Biodeterior. Biodegradation, 84*, 204-210.
[http://dx.doi.org/10.1016/j.ibiod.2012.03.001]

Shah, A.A., Hasan, F., Akhter, J.I., Hameed, A., Ahmed, S. (2008). Degradation of polyurethane by novel bacterial consortium isolated from soil. *Ann. Microbiol., 58*(3), 381-386.
[http://dx.doi.org/10.1007/BF03175532]

Singh, B., Sharma, N. (2008). Mechanistic implications of plastic degradation. *Polym. Degrad. Stabil., 93*(3), 561-584.
[http://dx.doi.org/10.1016/j.polymdegradstab.2007.11.008]

Thomas, R.T., Nair, V., Sandhyarani, N. (2013). TiO_2 nanoparticle assisted solid phase photocatalytic degradation of polythene film: A mechanistic investigation. *Colloids Surf. A Physicochem. Eng. Asp., 422*, 1-9.
[http://dx.doi.org/10.1016/j.colsurfa.2013.01.017]

Urbanek, A.K., Rymowicz, W., Mirończuk, A.M. (2018). Degradation of plastics and plastic-degrading bacteria in cold marine habitats. *Appl. Microbiol. Biotechnol., 102*(18), 7669-7678.
[http://dx.doi.org/10.1007/s00253-018-9195-y] [PMID: 29992436]

Wilkes, R.A., Aristilde, L. (2017). Degradation and metabolism of synthetic plastics and associated products by *Pseudomonas* sp.: capabilities and challenges. *J. Appl. Microbiol., 123*(3), 582-593.
[http://dx.doi.org/10.1111/jam.13472] [PMID: 28419654]

Yamada-Onodera, K., Mukumoto, H., Katsuyaya, Y., Saiganji, A., Tani, Y. (2001). Degradation of polyethylene by a fungus, *Penicillium simplicissimum* YK. *Polym. Degrad. Stabil., 72*(2), 323-327.
[http://dx.doi.org/10.1016/S0141-3910(01)00027-1]

Yang, J., Yang, Y., Wu, W.M., Zhao, J., Jiang, L. (2014). Evidence of polyethylene biodegradation by bacterial strains from the guts of plastic-eating waxworms. *Environ. Sci. Technol., 48*(23), 13776-13784.
[http://dx.doi.org/10.1021/es504038a] [PMID: 25384056]

Yang, Y., Yang, J., Wu, W.M., Zhao, J., Song, Y., Gao, L., Yang, R., Jiang, L. (2015). Biodegradation and mineralisation of polystyrene by plastic-eating mealworms: Part 1. Chemical and physical characterisation and isotopic tests. *Environ. Sci. Technol., 49*(20), 12080-12086.
[http://dx.doi.org/10.1021/acs.est.5b02661] [PMID: 26390034]

Yoshida, S., Hiraga, K., Takehana, T., Taniguchi, I., Yamaji, H., Maeda, Y., Toyohara, K., Miyamoto, K., Kimura, Y., Oda, K. (2016). A bacterium that degrades and assimilates poly(ethylene terephthalate). *Science, 351*(6278), 1196-1199.
[http://dx.doi.org/10.1126/science.aad6359] [PMID: 26965627]

Zhang, DJ (2019). biodegradation of polyethene microplastic particles by the fungus *Aspergillus* flavus from the guts of the wax moth *Galleria mellonella. Sci. Total Environ.*
[http://dx.doi.org/10.1016/j.scitotenv.2019.135931] [PMID: 31830656]

SUBJECT INDEX

www.ingramcontent.com/pod-product-compliance
Lightning Source LLC
Chambersburg PA
CBHW041720210326
41598CB00007B/728